THE FEAST OF
THE OLIVE

Aris Books
Berkeley • Los Angeles

THE FEAST OF THE OLIVE

Cooking with Olives & Olive Oil

By Maggie Blyth Klein

Illustrated by the author

The recipe "Pane con Olive" is from *The Fine Art of Italian Cooking* by Giuliano Bugialli with permission from Quadrangle/The New York Times Book Co., New York, 1977.

The recipe "Baked Rockfish Veracruz Style" is from *The California Seafood Cook-book* by Isaac Cronin, Jay Harlow, and Paul Johnson with permission from Aris Books, Berkeley, California, 1983.

Published by L. John Harris
Project Editor: S. Irene Virbila
Production Editor: Robin Cowan
Recipe Tester: Jay Harlow
Contributing Editor: Charles Perry
Book and jacket design by Linda Lane and Kajun Graphics
Printing and binding by Maple-Vail Book Manufacturing Co.
The binding on this book has been Smyth sewn for strength.

Library of Congress Cataloging in Publication Data

Klein, Maggie Blyth, 1947–
The feast of the olive.

Bibliography: p.
Includes index.
1. Cookery (Olives)   2. Cookery (Olive oil)
I. Title.
TX813.04K54   1983   641.6'463   83-15457
ISBN 0-943186-09-9
ISBN 0-943186-08-0 (pbk.)

Aris Books
Harris Publishing Company, Inc.
1635 Channing Way
Berkeley, California 94703

First Printing August 1983
2 4 6 8 9 7 5 3
Manufactured in the United States of America

FOR ROB

ACKNOWLEDGMENTS

My heartfelt thanks to my friends in Tuscany, Piero and Lorenza Stucchi-Prinetti, for their generous and warm hospitality. Thanks also to Maurizio Castelli for his teachings, and to Virginia Nicholls for her trail blazing. Particular thanks to John Meis for withholding nothing—neither congeniality, nor anecdotes, nor marvelous meals.

My gratitude to Marion Blyth, Frank and Mary Robertson, Charles Perry, Nancy Friedman, and Eugenio Pozzolini for making available the results of their research.

Thanks to Pat Darrow, Anzonini del Puerto, Julie Smith, Peggy Anne Davis, Isaac Cronin, and Ann Walker for contributing their wonderful recipes; and to Pam Fabry for her invaluable artistic wit and advice.

My warmest thanks to my publisher, L. John Harris, who had the initial idea for a book on olives and olive oil, and whose contributions—culinary, aesthetic, and editorial—helped give shape to the book. To Robin Cowan, excellent editor and patient coordinator; Sherry Virbila, editorial and aesthetic advisor; Jay Harlow, fine chef and food advisor. And thanks to the special people at Aris—Isaac Cronin, Mimi Luebberman, Debbie Bruner, *et al.*

My appreciation to Linda Lane and Jeanne Jambu for their refined sensibilities and beautiful book design.

My gratitude to my dear friends who helped me with the book by eating.

And eternal indebtedness to the indomitable Rob Klein, and my wonderful parents and brother, Richard.

PREFACE

*O*lives and olive oil are so fundamental to the history of the human diet that one wonders why no previous book has celebrated their virtues. The rapidly growing appreciation for fine foods in the United States has made such a book inevitable.

Enter Maggie Blyth Klein, a friend from my college days and one of the best kept secrets of the San Francisco Bay Area's culinary universe. Since the mid '60s, when we attended a "gourmet club" with our friends, Maggie has been devoted to fine food. Dining "Chez Maggie" is truly a grand gastronomic event, and cooking from her book is the next best thing.

As senior editor of the University of California's Agricultural Sciences Publications, Maggie Klein's technical knowledge and research skills broaden and give depth to the issues implicit in her subject matter. Accordingly, *The Feast of the Olive* is more than a marvelous collection of recipes; it treats the olive and olive oil in their botanical, medicinal, historical, religious, and literary totality.

From the olive oil factories of Tuscany, to the plush up-town delis of Manhattan, to the olive groves and agricultural research facilities of California, Maggie has tasted and tested, watched and listened. She is an investigative reporter with the skills of a chef, the talents of an artist, and the curiosity of an academic, uncovering the age-old secrets of a neglected staple.

Olives and olive oil, like few other foods, invoke an atmosphere we in America seek to recreate in our cooking: A sunny, southern realm—be it Provence, Tuscany, or a Mediterranean seashore. A world of robust, earthy tastes and smells. Jeremiah Tower, one of California's finest chefs, sent us a word image which perfectly conveys the vision of Maggie's Mediterranean olive feast. Told to Jeremiah by the great cookbook writer Elizabeth David, the scene involves Ms. David and the legendary Norman Douglas, author of the '40s novel, *South Wind.*

> A mid-summer afternoon, the restaurant high up on the cliffs at Capri, passing down an arbor passage hanging with grapes and jasmine, then through a tiny courtyard of flower-bearing orange trees, bees everywhere, to sit at a table

cooled by high-up-from-the-Mediterranean breezes, to a glass of simple, local white wine, feeling the cold liquid sliding down a dry and slightly dusty throat...

A bowl of olives, a carafe of green oil, some peasant bread, salt. And, of course, some of the most scandalous conversation in Europe.

Thank you Maggie for bringing so many facets of this feast to life through your wonderful book.

L. John Harris
Publisher, Aris Books

Contents

INTRODUCTION 14

THE HISTORY OF THE OLIVE 20

OLIVE OIL 30

THE OLIVE OIL OF TUSCANY 44

OLIVES AT THE TABLE 52

AN OLIVE GLOSSARY 58

RECIPES 67

ABOUT THE INGREDIENTS 68

LITTLE RECIPES 72

HOME-CURED OLIVES

HORS D'OEUVRES

BREADS, PIZZAS, AND SANDWICHES

FIRST COURSES AND SAUCES

THE OLIVE IN AMERICA 174

THE BOTANICAL AND HORTICULTURAL OLIVE 186

GROWING YOUR OWN 193

THE NUTRITIVE OLIVE — FACT AND LORE 196

THE OLIVE IN ART AND LITERATURE 206

THE FEAST OF
THE OLIVE

INTRODUCTION

I was born in California. Special dinners at my family's house always included a dish of green and "black-ripe" California olives. I would steal both kinds from the dining room table as the setting sun filled the room and my mother was upstairs getting dressed for company. My rearrangement of the olives was a perfect coverup for the theft, I'm sure. I wonder if at that age, I would have loved shriveled Moroccan olives or cracked Sicilians as much. But I was in Los Angeles, not in the Lozère.

And ever since I began to cook, my taste preferences have been Mediterranean. I never craved sushi or gravlax though I liked them just fine. No doubt my mother thought I would give up my aberrant liking of olive oil. Instead, here's a book called The Feast of the Olive. *Mother can still take or leave olive oil.*

If the cuisine of the Mediterranean could be characterized by any one flavor—one element that would change it utterly were it never to have existed—that flavor would have to be olive oil. Without it, the Berber shepherd, the gypsy in Andalucia, the housewife in Provence, the Tuscan vintner, the fishmonger in Calabria would never have been able to enjoy the wonderful meals they and their ancestors have been eating for hundreds of years. Food prepared with the singular, fruity, and glorious taste of a good olive oil, accompanied by its old companion, garlic, is enough to transport anyone to some sunny clime close by the Mediterranean.

The importance of olives and their oil is not merely gastronomic: Olive oil has represented holiness, healthfulness, and plenty and has illuminated temples and homes for thousands of years. Olive oil comes from the most noble of trees, and from the most ignoble of fruits. It is astonishing that the ancients, with their limited knowledge of machinery and technology, discovered not only how to extract the oil from such a bitter and unpromising source, but also how to do so without damaging or adulterating the delicious liquid.

An ever-growing array of handsome jars and cans of olive oil is filling our stores. Can we cooks who are anxious to create the flavors of the Mediterranean expect to find among them oils that compare with or surpass the quality of the oils made for Mediterranean cooks long ago? With a little knowledge, we can. It is unfortunate that the simple methods that produce the purest, least modified olive oils are being replaced more and more by industrial processes. Consequently, the words *extra virgin* and *pure* do not always mean what they used to. To help you choose the proper oils, this book includes an outline of the processes by which olive oil is obtained, the factors that lessen or heighten the quality of an oil, and the current labeling practices and legislation pertaining to them.*

As difficult as it is to unravel the new mysteries about olive oil, it is a simple and more intuitive undertaking to become familiar with and appreciative of the tree from which olive oil is derived. Aldous Huxley described olive trees as "numinous." Their great height and breadth, the rock-breaking strength of their tremendous, hidden roots, the trunks through which life flows skyward to the animated leaves, do suggest divinity. And, to my mind, the olive is the best of them, the most benignly numinous of trees. Born with civilization itself, the cultivated olive tree provides the universal symbol of peace, the olive branch, and can pacify even the most restless of us if we sit in a warm, timeless olive grove. It offers gifts of great variety: light (how many wicks in bowls of olive oil have been lighted throughout history?); heat (there are few better-burning woods than olive wood**); year-round shelter from the sun and rain (the olive tree is an evergreen); and food (green, brown, purple, and black cured fruit, and the sumptuous green-gold liquid). It is tenacious and undemanding, thriving in the most inhospitable terrains. And it seems to live forever. No wonder the Hebrews, the Moslems, the ancient Greeks, and the Christians made so much of this tree, its fruit and oil.

A fascinating study could be based on tracing the interconnections of Mediterranean peoples through their culinary uses of the olive and olive oil. For example, were country bread, olive oil, and garlic combined in one geographic location and then taken systematically all

---

*Because such standards are in a state of flux, and a certain amount of industry self-regulation is beginning to take place, guidelines for buying oil will change somewhat from one year to the next.

**37.3 million BTU/cord, compared with 28.2 BTU/cord for the white oak.

Hook lamp — very humble but very useful for many centuries.

over the Mediterranean, or does that combination of ingredients go together so naturally that it sprang up independently in various places? And what could be postulated about the famous north/south split that separates Spain, the Maghreb, France, Portugal, and most of Italy (those countries where the olive fruit is used extensively in traditional dishes) from Greece, Syria, Turkey, Egypt, and Yugoslavia (where olives are eaten in great quantity but only as small meals in themselves or as hors d'oeuvres and garnishes, but never cooked in dishes)?

That the olive is a precious part of the life of the Mediterranean is reflected by its abundant use in place names throughout the Mediterranean and around the world. From the Semitic word for olive, *zeit*, we have Zejtun in Malta; Wadi Zeit in Jordan; Mt. Zeit, a range in Egypt; Zeytindag, a mountain in Turkey; Zeytinburnu, "olive point," also in Turkey; El Zeitun, a village in Egypt; and so on. Derived from the Greek *elea* are the olive-infused place names in Europe and the

Western Hemisphere. Ela is a cape and river in Cyprus; Elaia, Elaik-horion, Elaiofiton, Elaion, and Elaiotopos are in Greece; towns named Olivet can be found in France, Michigan, New Jersey, and South Dakota; Olivone is in Switzerland; towns named Oliva can be found in Argentina, Spain, and the Canaries; Oliveira dos Brejinhos is in Brazil; Olival and Olivais are in Portugal; Oliva is a mountain in Chile; the Mount of Olives is in Jerusalem; the Olivares River, a peak called Olivares de Júcar, and towns called Olivares can be found in Spain; Cerro de Olivares is a peak in Chile; Olivia is the name of towns in Minnesota and Texas.

It is probably an accurate assumption that the olive gained such importance because of its indispensable oil. It is difficult to imagine that the Mount of Olives was named after a tree because its fruit made a pleasing pickle. But in spite of the cured olive's being overshad-owed by its sister product, it nevertheless is an important and irresis-tible creation. A bowl of *olives cassées*, placed before guests while their hostess concentrates on a *canard rôti* in her Languedoc kitchen; the purple, pointed Kalamata flattering a morsel of feta cheese at a picnic in Greece; shiny, black Moroccan olives, swimming in a juicy tagine in Fez—all seem as indigenous to those settings as lichen on a rock. You would be hard pressed to find a recipe for small game in Sicily that did not include olives, a platter of thrushes not replete with olives in Provence, or a breakfast table in Egypt or Syria without olives, cheese, and bread.

More and more, the various olives of France, Italy, and Greece are finding their way into delicatessens and specialty shops, as well as large markets, here in the United States. We now have the special pleasure of tasting olives shipped in their mother brine—some unpas-teurized—made according to some old-country recipe, and packed in bulk straight from the barrel in their country of origin. Many Ameri-cans are tasting them for the first time and learning to prefer them to the milder "black-ripe" olives of California. Certain olives—the dry-cured blacks, for example—are still too bitter for the long-dormant taste buds of many of us. But the day will come when we will take the more pungent olives for granted as we buy cracked green water-cured olives, for example, to pique the appetites of our dinner guests.

One would think that olive oil and olives would have made the trip to America long ago, with the first Mediterraneans who settled here. But circumstances dictated otherwise. Olive oil was expensive to import and many unscrupulous dealers sold bad-tasting, adulterated,

Olive picking in Catalunya.

or even dangerous substances as "pure olive oil." The use of olive oil did not catch on, and though it remained a staple of the Mediterranean immigrants to America, its use did not survive with succeeding generations.

There was no domestic supply of olive products until the early twentieth century, because the olive grows well only in California (and a few other western states), and because California was late in becoming an agricultural exporter. When California did produce oils it was for a small market, and many of the California oil producers went out of business after World War II when they were undersold by importers.

One would think, too, that in the American melting pot the full-bodied flavors of the Mediterranean would overcome the milder flavors of American cookery rather than the other way around. But such was not the case. When Emerson, writing in the late 1880s, was looking for something exotic and distasteful to compare with life at sea, he chose the olive. Apparently he could get used to neither. Nor was he alone. By the time California started selling cured olives commercially, the American palate was too sleepy for anything but mild "black-ripe" olives, the type still widely eaten today. But our tastes have begun to change, and a growing number of small California olive producers are selling a wider variety of olives.

Americans also grew used to bland, predictable, and inexpensive cottonseed oil. When we finally switched to other oils it was for health's sake, and corn and other mild oils were chosen. In our fitness-conscious age, it has now been determined that olive oil is a neutral oil in the cholesterol controversy—it neither promotes nor decreases cholesterol buildup. Perhaps this factor, combined with America's rising interest in so-called gourmet cooking, has encouraged a considerable number of Americans to opt for olive oil, both foreign and domestic. So far, novices tend to equate any pungency with quality, or, alternatively, blandness with subtlety. But when we shop and compare, cook and taste, our extra virgin palates will become, I dare say, quite refined.

The History of the Olive

The whole Mediterranean,
the sculpture, the palms,
the gold beads, the bearded
heroes, the wine, the ideas,
the ships, the moonlight,
the winged gorgons, the bronze
men, the philosophers—
all of it seems to rise in the
sour, pungent taste of these
black olives between the teeth.
A taste older than meat,
older than wine. A taste
as old as cold water.

Lawrence Durrell
Prospero's Cell

$\mathcal{T}$he olive tree, *Olea europaea*, has a history almost as long as the history of western civilization: Its development was one of civilized man's first accomplishments. How odd that no philosopher or historian has had the good sense to solve the considerable problem of defining *civilization* by saying that the first civilized man was the one who developed the olive, that civilization is an advanced state of human society made possible by the olive. Since its development, the olive has been a symbol of peace and of life's bounty, the subject of mythology, a source of light, and the very flavor of the Mediterranean.

The wild olive, or oleaster, grows in most of the countries of the Mediterranean and, in its numerous varieties, in southern and eastern Africa, southwest Asia, and other areas. It is an unimpressive, straggly plant, which bears tiny, inedible fruit, and has little resemblance to the magnificent *Olea europaea*.

The century in which husbandmen first cultivated *O. europaea*, and whether it was developed from the oleaster, remain mysteries. It may have been first cultivated independently in two places, Crete and Syria. The earlier development is generally conjectured to have been in Syria, probably by a Semitic tribe, perhaps as long as five thousand years ago. The Egyptians called the olive by a Semitic name, suggesting they were introduced to it by the people of the Levant (the area bordering on the eastern Mediterranean).

CANAAN

By Biblical times, the olive was already growing in great abundance in the land of Canaan, for there are innumerable Biblical references to its harvest and instructions for its cultivation. In Canaan olives and olive oil were used from remotest antiquity for food, cooking, oil, medicine, salve, lamp fuel, and, most important, anointment. Exodus 30:22–33 reads: "Moreover the Lord spake unto Moses, saying, Take thou also unto thee principal spices . . . and . . . olive oil. . . .

And thou shalt make it an oil of holy ointment, an ointment compound after the art of the apothecary: it shall be a holy anointing oil. And thou shalt anoint the tabernacle of the congregation therewith, and the ark of the testimony. . . . And thou shalt speak unto the children of Israel, saying, This shall be a holy anointing oil unto me throughout your generations. Upon man's flesh shall it not be poured, neither shall ye make any other like it, after the composition of it: it is holy, and it shall be holy unto you. . . . Whosoever putteth any of it upon a stranger, shall even be cut off from his people." Note that the text specifies olive oil for the making of the unction that makes one holy. The Catholic, Orthodox, and other churches still use perfumed oil made with olive oil for important rituals. *Chrism*, from which the word Christ is derived, means "anointed" and therefore "holy."

The olive was a symbol of safety and a harbinger of plenty for Noah when the ". . . dove came in to him in the evening; and, lo, in her mouth was an olive leaf pluckt off." Jerusalem was founded at the foot of the Mount of Olives; Gethsemane, the garden outside of Jerusalem,

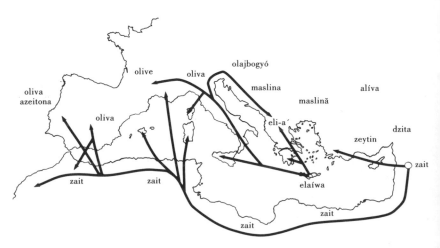

The spread of olive cultivation after the development of the domesticated olive in Crete — where it was called elaíwa — and in Syria — where it was called zait by the Semites.

means "oil press" (*gathsemen*). God blessed mankind, says the Old Testament, with "wine that maketh glad the heart of man, and oil to make his face to shine" (Psalms 104:15).

EGYPT

Egypt was importing two olive oils in the third millenium B.C., one from Syria, the other from the west side of the Nile Delta. Egypt did grow her own olives, but they were known only for their flesh, the oil apparently being of an inferior sort. Egyptian depictions of olive trees date from 2000 B.C., and three-thousand-year-old mummies are preserved with oil, among other things, and adorned with olive branches. Cured olives were left in the tombs of the Pharoahs for food in the afterlife.

In Roman times, the Egyptian olive groves of the oasis of Fayyum were quite famous for their olives. But the cultivation of olives never became as widespread in Egypt as it did in the North African countries to the west.

THE MINOANS

Early representations of olive twigs in Minoan art suggest that olives were being grown on Crete as long ago as 2500 B.C. Because there is no indication of contact between the Levant and Crete at that time, one might conjecture that the two areas developed olive cultivation independently. The derivations from the two names for olive would support that theory: The Semitic name *zayt* became *zayith* (Hebrew), *zaita* (Aramaic), *zait* (Arabic), and *dzita* (Armenian), and spread south and west through the Levant and the Maghreb (northwestern Africa); the Cretan word *elaíwa* became *elaía* in classical Greek. It appeared in Latin as *oliva*, and then in Celtic languages (as *olew*, for example, in Welsh), and spread north and west throughout Europe.

A remarkable feature at the palace of Knossos was the Room of the Olive Press, with open pipes that fed into storage vats. The great jars—amphorae—in which the oil was stored exist to this day. Olive oil became an important trade commodity for the Minoans, who probably shipped olive cuttings to Greece, the north coast of Africa, and other locations in the very ships that carried the oil. Samos, on the Minoan sea route to the Gulf of Smyrna, means "planted with olives."

GREECE

Greek mythology describes how the precious olive tree was brought to Greece: Zeus had promised to give Attica to the god or goddess who made the most useful invention. Between the horse that Poseidon produced—beautiful, rapid, and capable of pulling heavy carriages and winning wars—and the olive tree that Athena produced—with oils that could illuminate the night, sooth wounds, and offer nourishment—Zeus chose the more peaceful invention and Athena became the goddess of Athens. A son of Poseidon tried to wrest the olive tree Athena had planted from the rock in which its roots were embedded, but wounded himself in committing the impious act and died. That rock is the Acropolis, where the original olive tree was guarded by soldiers. Aristides said the center of Greece is Attica; the center of Attica, Athens; the center of Athens, the Acropolis. Taking things a step further, the center of the Acropolis must be the sacred olive tree.

When olive oil was first imported into Greece, in the pre-Homeric era, it was used as a costly anointment and as a component of perfumes. Its use in lamps became widespread somewhat later, and it was later still that olive oil became essential to Greek cookery. The olive "berry," however—pickled, brined, and highly seasoned—is referred to many times by Homer in *The Odyssey*, as is the olive tree. It is assumed, therefore, that cuttings were first introduced into Greece around the tenth century B.C., at which time it was realized that the calcareous soil of Greece, which had proved inhospitable to many other crops, was well suited to the undemanding olive tree.

By the seventh century B.C., olive orchards were well established on the limestone hills of Attica. By the beginning of the sixth century B.C., the great statesman Solon had enacted laws protecting the olive trees of Athens. Inspectors checked monthly to see that no trees had been cut down; anyone convicted of doing so was executed. Herodotus, in the fifth century B.C., described Athens as a vast center of Greek olive culture. Oil was produced in such abundance that it became one of the major exports. Its sale was profitable indeed. Greece also marketed abroad her oil lamps, made of molded terracotta, and spread her techniques for olive cultivation and olive curing over the whole of the Mediterranean. (The area around Aix-en-Provence, for example, was one place where Greek cuttings took root.)

The olive branch became a symbol for the Greeks: Olive-leaf wreaths adorned the brows of accomplished soldiers as well as scholars; tombs unearthed in the 1930s revealed wreaths of gold olive leaves. A frieze on the Parthenon depicts Athena's olive tree, and victorious athletes receiving olive branches and bowls of olive oil.

The olive tree was so revered in Greece that it is said only virgins and chaste men were allowed to cultivate it. An oath of chastity was demanded of those who took part in the harvesting.

Ancient navigators protected themselves from the wrath of Poseidon by placing an olive branch between the hands of an image of their tutelary god. Athena's olive, having defeated Poseidon once, could only strengthen their god's protective power.

A special wreath of olive branches, wrapped with wool, was carried by singing boys during harvest festivals, and was later hung on house doors. Olive branches dipped in purifying water were used in funeral ceremonies.

ROME

The olive was a symbol for the Romans just as it had been for the Greeks. They too mingled its leaves in their triumphal crowns. And "Wine within and oil without" became the happy formula for Roman well-being. Having been taught the use of oil in gymnasiums by the Greeks, the Romans took it to extremes. Pliny complained that scrapings of oil and sweat from the great athletes were being sold by the gymnasiums for 60,000 *sesterces* for use in medical plasters and emollients.

With their ascendancy, the Romans continued developing all aspects of olive cultivation and curing, and olive oil production. After they invented the screw press, the Romans began producing olive oil as it was to be produced for the next two thousand years: Olives were crushed by two hemispherical pieces of stone attached to an axle set atop a post in the middle of a circular trough; as the stones turned, the olives were crushed. The oil was then extracted by a press (*torcular*)—either a screw type, or a more primitive type, including one with two logs against which wedges were hammered to exert pressure. (That ancient method is still in use in Tunisia.) After the oil was extracted, the water from the olives' flesh and sediment settled out in holding tanks; then the oil was stored in large, clean urns. (The twentieth-century centrifuge, which separates the water from the oil, is the only new development in the production of some of today's finest olive oils.)

Obtaining virgin olive oil in the XVIth century – from an old engraving. (after a reproduction in Els olis de Catalunya.)

As olive oil became more widely used in cooking, the standards to judge it by were established. Cato explained that for Romans, the finest oil was "bitter oil"—that made from "white" olives (small, immature olives) picked in September. Oil from green, December-picked olives (*oleum viride*) was good-quality table oil. The oil pressed from black olives was used at the table by common people, and for lamps and anointment. The fruit for fine table oil was held for only a short time after harvesting lest it become rancid, then the pulp was subjected to moderate force. The first pressing produced the finest oil; the product lost quality as additional force was applied during the second and third pressings. The lowest quality oil was that obtained from fruit that had rotted, been damaged by vermin, or was so dirty it had to be washed with hot water.

The area that grew the best oil olives was thought to be Venafra in Samnium (Campania). Rome imported oil from Spain, Tunisia, and Istria (in Greece), often as payment for taxes, although its own oil was considered the best. However, a fourth- to second-century B.C. cookbook entitled *De Re Coquinaria* gives a recipe for faking the high-priced Liburnian oil from south of Istria: in Spanish oil, steep pounded helenium, cyprus root, fresh bay leaves, and salt.

In Rome, pickled and cured olives were eaten in large quantities. They were particularly enjoyed as hors d'oeuvres and were thought by some to be an aphrodisiac. Olives have been found still preserved in the ruins of Pompeii. The recipes from ancient Rome are as sophisticated and delicious sounding as any found today. The Romans cured olives at all stages of ripeness. Often, the bitter glucosides were leached by soaking the olives in a wood ash paste. The green olive was preserved in salt brine; or it was mashed and soaked in frequently changed water baths, then soaked in a salt brine flavored with vinegar, fennel, and other seasonings. Sometimes sweet wine or honey was substituted for the vinegar. Half-ripe olives were picked with their stalks and allowed to cure in olive oil. Ripe olives were sprinkled with salt and held for five days, after which time the salt was shaken off and the olives dried in the sun. Sometimes pitted ripe olives were cured in jars filled with oil and coriander, cumin, fennel, rue, and mint. Such olives were referred to as *oil salad* and were eaten with cheese.

There were also olive cakes called *sampsa*, made from pressed olive pulp flavored with salt, cumin, anise, fennel, and olive oil. The rustic dish was hawked on the streets by strolling vendors and, supposedly, would last two months before going rancid.

Some recipes from *De Re Coquinaria*:

— A sauce for boiled chicken made of pepper, cumin, thyme, fennel, mint, rue, asafoetida, vinegar, dates, honey, *liquamen* (a fish sauce), olive oil, and salted olives.

— A stuffing for boiled chicken made of chopped fresh olives stuffed into a chicken, which is then sewn closed. The olive stuffing is removed when the bird is done.

— A vegetable dish of boiled cabbage arranged in a shallow pan and dressed with *liquamen*, oil, wine, cumin, pepper, chopped leeks, caraway, coriander, and olives.

OTHER PLACES

By the time Rome fell, the olive tree was flourishing in all the temperate climates within the Roman empire. Olive orchards were so common in Spain that Hadrian adopted the olive branch to symbolize Roman Hispania. The Moors found the tree thriving in Catalunya. Olive culture was at its height in Tunisia during the Roman occupation. (Rome taxed the province 300,000 gallons of olive oil yearly, and by the time the Arabs conquered the territory, olive orchards covered over two million acres.)

A curious pattern began to develop in the cuisines of the Mediterranean. In Greece and in the countries around the Levant, the fruit of the olive came to be used only as an appetizer, albeit in great abundance. But in the Maghreb and the northwestern parts of the Mediterranean, olives came to be added freely to dishes; in Morocco, in fact, many pungent stews, called *tagines*, are described as "smothered in olives." Oddly, the oil itself is rarely seen in Moroccan cuisine, yet it is used quite a lot in the Levant, as it is, of course, in Italy, Greece, Spain, and Portugal.

To give you an idea of how olives and olive oil were used, here are some recipes collected by Charles Perry, a foremost scholar on the cuisines of the world and their history:

♦ From a thirteenth-century Spanish Arab recipe: Clean a partridge and put it in a pot with salt, coriander, pepper, chick peas, cinnamon, and other spices, with two spoonfuls of oil, water, lemon leaves, and fennel stalks. Make small meatballs from the breast meat of the partridge and cook in the same pot. After it comes to a boil the third time, take the partridge out and put on a board. Take chard stalks cut in quarters, tie them in a bundle, and throw them in the pot with olives. Fry pieces of cheese and add them, and remove two or three meatballs, pound them in a mortar, and mix with beaten egg white. Cook the egg yolks in the stew, cover with the egg white mixture, and cook until done. Serve with the partridge, meatballs, egg yolks, pieces of cheese, and olives.

♦ From a thirteenth-century Syrian recipe: Take olives from Palmyra (black for preference), remove the pits, and "sift," mixing with cardamom and ground walnuts. Sprinkle with coriander, toasted walnuts, and salted lemon, knead together, and put in a jar.

♦ From a thirteenth-century Iraqi recipe: Take ripe or green olives (black are better) and crush and salt them. Turn them over every day until their bitterness disappears, then put them on a tray of woven sticks for a day and a night until dry. Pound garlic and dry thyme with an equal weight of walnuts. Put the mixture on a low fire, and put the tray of olives on the same stone in an oven, close the door, and leave a whole day. Stir several times so that the aromatics circulate in them. Take out and season with sesame oil, crushed walnuts, toasted sesame seeds, garlic, and thyme.

In spite of the importance of olives and olive oil around the Mediterranean, it was not until 1560 that olive cuttings were taken once again a great distance, this time to Peru by the Spaniards. In the 1700s Franciscan fathers brought the olive to Mexico, and then north into California; but it was not until the late 1800s that California commercially developed olive cultivation. Thomas Jefferson was one of many horticulturists who tried to grow the worthy tree at home, but the climate at Monticello—like the climate of England, where similar attempts had been made—was inhospitable to the olive. The olive was introduced successfully in Australia over a hundred years ago, and was taken to South Africa in 1903 by an Italian named Costa. New plantings are currently being tested there. But most of the world's olives are still grown near the Mediterranean, where the acquired taste for olives has been with its growers for hundreds of generations, and might by now—one never knows—have found its way into the gene pool.

OTHER USES

Records of the commercial role of olive oil through the Middle Ages until modern times show that once the olive tree was established, its products became essential throughout the Mediterranean and the world. Olive oil illuminated Mediterranean homes well into the nineteenth century (and in some places into the twentieth) when it was replaced by coal and petroleum oils. Olive oil was used for making soap all around the Mediterranean and is still a common fat used in the saponification (or soap making) process. It was a prime lubricant for the machines that powered the industrial revolution—two thousand years after it was first used as axle grease by the Romans. And jewelers to this day polish their diamonds with fine olive oil.

Olive Oil

Olive oil is found, not made.

Professor Cupari

There's more to this
than meets the eye.

M. Klein

*N*o matter how much one knows about olive oil, the most valuable aid to purchasing good oil is a knowledgeable merchant whose integrity is beyond question. That merchant will know the makers of the oils he stocks and the circumstances under which the oils are made, and will store the oils properly. Once this merchant is found, it is up to the consumer to taste and compare, and to buy oils in small quantities until he finds one that pleases his palate. (Some fine oils are now being sold in 8.4-ounce bottles.) Even then he must not limit himself to one label, for oils vary according to the kind of year it has been for olive production, and the kind of olives that are available to the oil manufacturer.

There are two basic rules for producing fine olive oils: The olives must be of the highest quality, and as little as possible must be done to them to extract their oil. For within the cellules of the red-ripe olive exist pockets of pure, delicious oil that is ready—without processing—to be savored.

The great olive oils of the world are produced on a small scale in the cooler climates of the olive-growing countries. Unfortunately, the small-scale producers of fine oils are finding it more and more difficult to compete with the large companies that sell refined oils. Production of true extra virgin olive oil—olive oil made from the first cold-pressing of quality olives—is said to have fallen 80 percent since the 1960s. Fewer and fewer farmers produce oil the old, careful way. Nevertheless, because of the expanding food consciousness in this country, importers are finding a good market for foreign foods, and the American consumer now has a good choice of first cold-pressed oils. Americans are getting a chance to taste the fruity and fragrant, light juice (and it is just that) that has been the cooking medium in much of the Mediterranean countryside for centuries. For the consumer, the difficult task is to differentiate between industrially produced, far inferior "extra virgin"-labeled oils and the traditional cold-pressed, truly extra virgin products.

How does one tell the good oils from the bland, refined oils, or from the poorly made, rancid tasting oils? Which of the less expensive oils should one buy for everyday cooking? Certainly price is no criterion by which to judge; the refined-oil makers are cultivating the gourmet market too, and often charge high prices for their oils, influencing inexperienced oil merchants or consumers to assume they have purchased the best money can buy. To the uneducated palate, the refined oils are a pleasant change from the vegetable oils to which Americans are accustomed. After all, we are recent graduates from cottonseed oil (widely used for decades and an oil that must be refined to be made comestible) to safflower, corn, and peanut oils. (Since most of these are refined as well, they are not much more flavorful.)

MAKING OLIVE OIL

What follows is a list of important factors that determine how good an oil will be once it reaches the table. Listing the factors may seem pedantic, but keep in mind that most everything that happens to an olive will be reflected in the flavor of its oil. The oil extracted from olives, like all fatty substances, retains the odors of almost everything near it. Those odors become flavors in the oil. (In fact, in the days of horse-powered oil mills, the farmers had to go to great lengths to ensure that the odor of the manure did not waft over the oil.)

Variety of olive. Olives vary as much according to variety as do apples or any other fruit. Some produce oil that is far superior to that of most others; some produce oil more plentifully. The large olives grown primarily as table olives in the hottest climates contain a small proportion of oil, which is often highly acidic. Their pits, being large and easily broken, are often crushed along with the flesh. The oil thus produced can have an undesirable, woody flavor.

Some varieties famous for their oil are the Arbequiña of northern Spain, the Frantoio of Tuscany, and the Nyons of Provence (although some would classify the Nyons as a table olive). Even in the hotter climates, distinctions are made—the Zorzaleños in Andalucia and the Mission in California are favored for their oil over the larger table olives with which they are grown.

Harvesting. The best oil is a blend of oil from olives harvested at the red-ripe stage and a smaller proportion of oil from green olives of a different variety. Some oil manufacturers add a few leaves to make an oil greener so it looks like that of pressings from the earlier harvest. Olives that are harvested late, when they are black, contain

The tin man from Pitigliano, in Tuscany, holding an oil can he made. (After a photo by S. Virbila.)

more oil than the red-ripe olives, but such oil is more acidic and generally of poorer quality. Much refined oil is made from the oilier, riper olives, then chemically de-acidified. And some virgin olive oils with fairly high acidity, but which are still comestible without refinement, are made from riper olives.

Olives can be harvested by hand or they can be beaten from the trees and collected in nets. Green fruit is so firmly attached to the tree that it must be plucked by hand. In some areas, olives are allowed to become so ripe they fall to the ground, to be gathered by hand, by machine, or in nets. Olives that are allowed to strike the ground are bruised and collect dirt, which hastens rancidity. Oil made from bruised olives is qualitatively different (i.e., chemically different) from oil made from undamaged fruit; an experienced taster can discern the difference immediately. However, oil producers who beat their olives from the trees and collect them in nylon nets claim that their oil is not affected by such treatment.

Most olive oil experts agree that hand-harvesting is the only way to make a superior oil. In some areas, such as the Spanish countryside, pickers use small hand tools that strip the branches of their fruit; in other areas the pickers use only their hands, wearing gloves or with palms taped for protection.

Methods of harvest vary greatly and affect the quality of the olive that is brought to the press; but unfortunately, the method of harvest is information to which the consumer probably will not be privy.

Storing the olives. Newly harvested olives are usually allowed to sit for at least a few days before they are pressed. As they rest, some of the water in their flesh evaporates, making processing easier. During the storage period it is essential that the olives are kept cool and that the internal temperature of the fruit never exceeds 86°F. Conscientious farmers spread the olives thin for ventilation; others pack their olives under ice to prevent the fermentation caused by heat.

Olives that have been stored too long before pressing will have a very high acid content and a foul odor and flavor, making chemical refining a necessity. The large producers of refined oils frequently store their olives for long periods of time, enabling them to stretch out the oil production process to keep the plant operating more months of the year during daytime hours. They generally rely on refinement to undo the damage done by lengthy storage.

Crushing. The crushing process is the same at most mills, whether large or small. Enormous granite stones, each weighing eight tons, crush the olives into pulp. Purists among the oil producers maintain that a softer stone produces less bruising; and conscientious mill operators have their stones chiseled every year to correct the smoothing that occurs during crushing. Rural farmers take their olives to the local mill, where they are crushed and pressed.

Pressing. In most small mills, the pulp produced by the millstones is layered on mats made of nylon or, as in the past, straw. The mats are stacked about four high, interspersed with metal disks, and then, under 300 to 400 tons of pressure, they are pressed in a screw or hydraulic press. The better mills do this without the addition of heat or water. The oil that is thus expressed is called the *first cold-pressing*. If the olives are not too ripe and have been treated well, the oil will easily qualify as extra virgin—a classification (see page 39) used in Spain, France, and Italy to denote less than 1 percent acidity. (In Catalunya, where labeling is even more strict, extra virgin olive oil is classified at under 0.5 percent acidity.) Because chemicals can produce zero acidity, however, the designations mean little unless accompanied by strict definitions for "cold-pressed." All fine European first cold-pressed oils are well under 1 percent acidity.

In addition to using screw and hydraulic presses, to which heat and water are applied, some larger factories (and smaller ones, in growing numbers) use large centrifugal presses that spin the crushed olives as heat and water are applied. (The centrifugal press is not to be confused with the small centrifuges that separate the water from the oil after pressing in small rural mills.) Such a process extracts more oil from the pulp but damages it in the process. It is to such presses that the pulp, or *sansa*, from the first pressing at the smaller mills is taken for further processing. The oil that results must be refined to be made comestible, and is then mixed with 5 to 10 percent extra virgin, or virgin, oil for flavor. Its label may read "*sansa* and olive oil."

Refinement. The highly acidic or foul-flavored oils produced from *sansa* or from badly damaged fruit are placed in a series of closed vessels. Chemicals with properties similar to those of gasoline (hexane, trichlorethylene, or carbon sulfide) extract the oil by percolation. In a steam-heated still, the solvent is completely recovered by condensation. The oils that result are pale, very fluid, nonacidic, and absolutely without flavor. Mixed with 5 to 10 percent virgin olive oil for flavor and color, they can be sold as pure or virgin oil or, if the acidity is low enough, as extra virgin oils. With a few exceptions— notably, the divine cold-pressed extra virgin oils from the small producers—such refined oil constitutes most of the olive oil available to American consumers.

Separating water from the oil. After the first (or in the case of the inferior oils, second or third) pressing, a percentage of the resulting fluid is water. That water is separated in a small centrifuge which spins at 1500 revolutions per minute, or it is allowed to settle out in a large holding tank. Most olive oil purists allow for this, the only new machine in the age-old process of making fine oils. But there are a few who contend that the centrifuge damages the oil, and that the best oils are made when the water is allowed to settle out.

Settling. The new oil is a cloudy liquid, full of minute particles of fruit in suspension. Most producers allow the oil to settle for a number of weeks, usually in huge earthen jars in a cool cellar. (The oil can be kept there almost indefinitely, until it is convenient for the producer to decant and bottle, or filter and bottle, the oil.) The new oil is ready to use, but is particularly peppery and fruity the first two months. A few producers bottle a small portion of the new, turbid oil and sell it as "new oil" to an elite few who are particularly fond of the potent liquid.

Filtration. It is not essential for good quality olive oil to be filtered. Some producers say that filtered oil keeps longer than unfiltered; some disagree and say filtration deprives the oil of flavor. Some feel that the clear oil is more aesthetically pleasing.

Fine oils are usually filtered through cotton; alkaline earth filters are used to reduce the acidity of poorer oils. The use of earth filtration can indicate an inferior product.

Variables. Just as some vintners buy grapes from other growers, many of the small olive farmers buy olives from other growers to augment their own production. This is especially true in years when crops have been damaged by pests in one area and not in another, or when a farmer waited too long to harvest and his crop is damaged by an unexpected and severe wet or cold spell.

The conscientious oil maker is as careful choosing the olives he buys as he is growing his own. Nevertheless, the different varieties, microclimates, and growing methods will produce oils with different characteristics. A producer might make a fruitier oil from the olives he has had to purchase one year than is characteristic of the oil usually made from his own olives.

Then too, different varieties of olives ripen at different times. If, as is common practice, a grower has three olive varieties in his orchards, he might harvest two-thirds of his crop red-ripe and one-third green. The proportions of ripe to green olives will have a marked effect on the oil, and will differ from year to year depending on the weather and the producer's olive sources.

Thus your preference for Brand X in January of one year may change to Brand Y in August or the following January. The best the consumer can do is know which brands of oils are made well—and there are currently more than a dozen superb oils being imported to the United States—and to continually compare the exquisite olive juices.

TASTING OLIVE OILS

Here's how an olive oil expert tastes oils: He pours a little of each oil into wine glasses and compares color and fluidity. He cannot determine fattiness by appearance, but he can usually tell which olive varieties make up each oil by its color; he is rarely stumped. Then he smells an oil, inhaling deeply. He sips a little, inhales while holding the oil between his soft palate and tongue, swishes it between his upper teeth and lip and on his gums, and inhales again. He spits it out and makes his declaration—"Ugh, fatty" or "Could be from Umbria, extra virgin, but made by a big company" or "Almondy. Very nice." A wine taster may taste at least twelve, sometimes as many as twenty, different wines at a tasting, but the limit even for an expert oil taster is six or seven. Eugenio Pozzolini of Dean and DeLuca Imports of New York is one such expert. He says that fine oils are tasted horizontally, that is, on the palate. A bad, acidic oil will be sensed vertically, in the throat, and leaves a bad aftertaste and a fatty coating in the mouth.

Begin a "tasting" by taking a little oil and rubbing it in the palms of your hands, then putting some on the back of your hand. Smell it. Note the aroma—whether or not it is fresh and fruity. If you are a novice, know as much about the oil you are tasting as you possibly can. Leave the blind tastings for later. You might start out with a good oil your merchant describes as mild—perhaps one mellowed by a year of good storage—and less of a jolt than, say, a peppery Tuscan extra virgin.

Olive oil is certainly a matter of preference, as much a subjective choice as with any condiment. Nevertheless, the most frequent contenders for best oil among most olive oil cognoscenti are the cold-pressed extra virgin oils of Provence, Liguria, and Tuscany. Connoisseurs who favor the Ligurian oil would say that it has more finesse than the oils of Tuscany, more life than the oils of Provence. Francophiles prefer the sweetness of the "lady oils," perhaps claiming they have a more highly developed and refined sense of taste than do the lovers of the Italian oils. Experts who favor the Tuscan oils would say the peppery aftertaste adds more life, and that the oils have more of a rustic taste and more body.

That the number of producers of such fine oils has fallen off so drastically is lamentable. The sad truth is that many European consumers, the major market for olive oil, would rather cook with the inexpensive refined oils of the huge olive oil companies than pay the price for the necessarily labor-intensive finer oils. The elaborate labeling laws do nothing to protect the manufacturer of carefully made oils, or the consumer who wishes to discern a well-made product by reading a label. The designations of oils from Spain, Italy, and France are based on acidity, and the acidity of an inferior product can be lowered quite easily with chemicals. There is nothing to stop the manufacturer of a de-acidified refined oil from calling it "extra virgin olive oil," so long as the oil is less than 1 percent acid.

Where does that leave American consumers who would like to have access to the best oils? We are, for the time being anyway, in the happy position of being a promising market for the small producers, if not their last hope. The consumption of extra virgin olive oils in the United States has been reported by one large importer to have risen 2000 percent since 1980. The American consumer is insisting on quality products; perhaps, as with the California wine industry, American olive oil producers will be inspired to improve their product to satisfy the increasingly discriminating American palate (see California Olive Oils, page 179). Some cold-pressed rural oils never before exported from Greece are expected soon in the American marketplace.

And what of the refined olive oils? Admittedly, they do not compare to the cold-pressed oils. But with virgin oil added to the deodorized oil, they make a much tastier everyday cooking medium than the wholly refined oils of most vegetables or seeds, and a more appropriate ingredient for some dishes. Because the expensive cold-pressed oils suffer when heated over 140°F, it is wasteful to use them in high-temperature cooking. I have enjoyed many fine meals prepared with refined olive oil.

The flavors of olive oil. Here are the words used most often by olive oil experts to describe the flavors and nuances of olive oil. *Almondy,* as the name suggests, is oil reminiscent of almonds. *Bland* oil is light, but with hardly any flavor; characteristic of refined oils. *Fatty* oil is high in fatty acids and is made from damaged, old, or over-ripe fruit, and leaves an unpleasant coating in the mouth. *Fluid* oil is not as thick as first-pressed oils; characteristic of refined oils. *Fruity* oil has the hearty flavor of olives; it is particularly apparent in oils with a high percentage of green fruit and in newly pressed oils. *Peppery,* a sensation somewhat like that produced by peppers, but not felt in the nose, is a quality found most pronounced in new and Tuscan oils. *Rustic* oil is hearty and flavorful and *sweet* oil is characteristic of lighter oils, such as those from Provence.

IDENTIFYING AND BUYING OLIVE OIL

♦ The labeling laws of countries with regulated denominations such as "extra virgin" (*vierge extra* in France, *extra vergine* in Italy, *virgen extra* in Spain) are not strict enough to ensure that an "extra virgin" oil is made from the first cold-pressing of olives. Although those three countries have the strictest regulations of any countries from which the United States imports olive oil, their nomenclature is based solely on acidity, an inadequate criterion of quality. Although the first cold-pressing of high-quality olives does produce oils of low acidity, an equally low, or lower, acidity can be honestly claimed on labels of chemically refined oils that have been totally de-acidified by alkaline solutions. If the classification "extra virgin" is determined solely by acidity, then a refined oil, with some first-pressed oil added for flavor and color, can be labeled "extra virgin" and catch the shopper off guard. The large companies are the ones that most frequently sell the bland, refined oils, but some smaller producers in the most prestigious areas of oil production can be equally guilty.

The following are the designations of Italy, France, and Spain:

Extra virgin olive oil
<1 percent acidity

Superfine (sometimes "fine") virgin olive oil
1.01 percent to 1.50 percent acidity

Fine (sometimes "regular") virgin olive oil
1.51 percent to 3 percent acidity

Virgin (sometimes "pure") olive oil
3.1 percent to 4 percent acidity

Lamp oil (not comestible)
>4 percent acidity

Olive oil
Rectified and refined lamp oil with
5 to 10 percent virgin olive oil added

Sansa and olive oil
Rectified and refined *sansa* oil (made from the pulp from first and second pressings) with 5 to 10 percent virgin olive oil added

Oils from Spain, France, and Italy that bear the words "extra virgin" or "superfine virgin" are either the first cold-pressing (the best quality oils) or *the rectified oils* (the last two classifications) *that qualify as "extra virgin" by having low acidity.*

♦ The regulations passed by the U.S. Food and Drug Administration in 1982 are better worded than the European laws and clearly state that "the name 'virgin olive oil' may be used only for the oil resulting from *the first pressing of the olives* [my emphasis] and which is suitable for human consumption without further processing." The regulation further states that blends of refined and virgin oils can be labeled "pure olive oil" but not "virgin." However, those regulations apply only to oils produced in this country. The federal government has not yet introduced definitive labeling criteria for more than those two grades of olive oil. According to the 1982 regulations, an American oil labeled "extra virgin" need not be different from an American oil labeled "virgin."

♦ Truth-in-labeling laws are well enforced in the United States. Adulterants in oils are easily detected in a laboratory by the use of saponification values, specific gravity values, and so on.

♦ The best insurance for the consumer is to develop the expertise to taste the quality of oils and, equally important, to find a knowledgeable merchant. Some merchants are now certifying their imported oils as being first cold-pressed.

♦ Generally speaking, the hotter climates produce fatty, lesser olive oils. However, all olive oils from a particular region, such as northern Italy, are not necessarily of the highest quality; nor are all Sicilian oils necessarily poor. There are pockets of cooler climates mixed in with the hot. Furthermore, a conscientious oil maker using a larger, fattier olive can make an enjoyable, honest product that might be less expensive than a deceptively labeled oil from a more prestigious, cooler district, manufactured by a corporation whose family name is not at stake.

♦ The olive oils of Greece, Portugal, and other countries with poor labeling laws will often be seen with such descriptions as "Extra extra virgin" on the labels. Such oils are in general of much lesser quality than fine olive oils, and can be inferior even to other inexpensive, more humbly labeled oils for everyday cooking. But there are a growing number of exceptions. The advice of a trusted merchant is a necessity.

♦ Although oils vary a great deal from year to year, and there are great years and not-so-great years, it would be silly to treat oil as if it were wine. Olive oil does not improve as it ages. Past the initial resting period, oil is best when it is newly pressed. Seeing the year of pressing on the label will enable you to avoid buying old oils and, if you have a good memory, to compare from year to year.

♦ Never buy an oil that has been stored in a sunny spot for any length of time, or that has a copper cast to its color.

♦ Test the labeling on your oils: Put a small quantity in a bowl and refrigerate it for a few days. If it forms small crystal-like structures it is probably extra virgin cold-pressed; if it becomes like butter it is probably "pit oil"; if it turns into a block, it is probably an industrially produced, chemically refined, "extra virgin" oil.

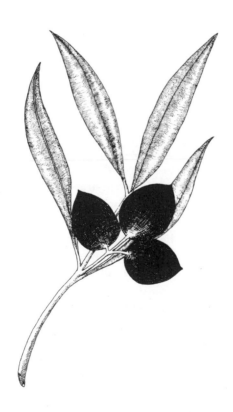

OLIVE OIL IN THE KITCHEN

Storing Oil

♦ Olive oil should be stored in glazed clay, very clean tin, stainless steel, or glass. Do not store it in copper or iron. Plastic will impart a flavor to the oil over time.

♦ Olive oil should be stored in a cool, dark place. Oil that has been affected by heat or light has a copper hue characteristic of oxidized oil. Well-stored oil can keep at least two years; some say it can keep indefinitely under perfect conditions. But remember that oil is at its best the first year after it is pressed, and at its most flavorful the first two months.

♦ Do not refrigerate olive oil. The best oil makers say to do so is like storing wine in a refrigerator; it interrupts the active life of the oil.

Cooking with Oil

♦ If you are using olive oil to cook with, it is pointless to use your best oil, as heating changes its character substantially. Find a good quality, less expensive oil that is to your taste for recipes that call for cooking with oil. I use a very pleasant, unfiltered extra virgin Spanish oil for cooking; others prefer California or Greek.

♦ The flavor of olive oil begins to change at 140°F. For that reason, fine olive oils should be added to dishes after they have been cooked, whenever possible. When barbecuing meat or fish, add the oil after cooking, unless oil is part of a marinade. Add olive oil to soups at the table. Olive oil is, in part, a condiment.

♦ Olive oil will begin to burn at 280°F, a low temperature compared with other oils, many of which can be heated to 365°F and higher. That means for frying foods, such as potatoes, whose texture depends on a very high temperature, it would be better to use another oil, such as peanut oil.

Reusing Oil

♦ Olive oil that is used for frying can be strained through cheesecloth, refrigerated, and used again, provided it has not been overheated. Do not keep olive oil that has been overheated, as overheating produces toxins. Oil used for frying fish should be reused only for frying fish, and then certainly not more than once.

Oil at the Table

♦ Of the oil cruets available, the best are those that do not drip. A modern, stainless steel cruet from Italy, which is being imported into the United States, collects the oil that would otherwise drip and funnels it back into the container. The long-spouted tin containers are pretty and more traditional, but sometimes messy. Glass cruets, or your favorite glass vessels, are fine, but must be stored away from the light.

Nutrition

♦ Olive oil is a monounsaturated oil and has no effect on the cholesterol level of the bloodstream. (And, as with all vegetable oils, it contains no cholesterol.) One tablespoonful of olive oil contains 125 calories. (For more on its nutritional qualities, see the chapter entitled "The Nutritive Olive.")

The Olive Oil of Tuscany

Why do we make oil?
Because we always have.

The farmers of Tuscany

For my taste, the best olive oils now available in the United States are from Tuscany. There are currently about ten top-notch Tuscan oils being imported from relatively small farms in the region. Such oils are characterized by their clean, strongly fruited flavor, their pepperiness, and their so-called rustic qualities. There is no doubt that besides Tuscany there are various locations around the Mediterranean that produce oils that are as good as, if perhaps different from, the Tuscan oils. A discussion of those oils would be superfluous because generally they are not exported—usually because production is in quantities sufficient only for the use of the families who produce them. By watching first hand the care with which the orchards of Tuscany are tended and how the work of the frantoios *(the oil mills) is carried out, it was easy for me to see why the olive oils of Tuscany are so exquisite.*

*T*hroughout the hills of Tuscany, and in particular those of Chianti—intermixed with the lovely, profuse vines that produce the Sangiovese grape—grow the small, bleak, twisted olive trees that produce the fine oils now finding their way into shops throughout the United States. Curious anthropologists come here to pick and sift among the Etruscan ruins. It is a land still, to a certain extent, owned by noble families whose lineages can be traced a thousand years or more. Here stretch miles of chestnut and umbrella pine forests, broken by rows of vines and clusters of well-tended olive trees, and dotted with fine old castles and the lovely *case coloniche* that were once in their domain.

In Florence, the old Tuscan center of commerce, history and the present live side by side; and in the medieval towns that crown the surrounding hills the same is true. Occasionally that affiliation is amicable, with modernity complementing antiquity, in some cases rescuing it, and sometimes tearing down the old orders that needed removing. Too often, of course, the present taints the past, adding a power line, an afternoon of brown sky, or a foreign conglomerate to fill the void left by a crumbled old fiefdom.

The old hamlet called Monti is a good case in point. It sits atop a hill; its old stone structures are inhabited by two families whose forebears' forebears' forebears lived there before them. They raise their

A Chianti olive oil takes first prize in Paris, 1927.

own poultry, rabbits, and vegetables, and look after the *vinsanto* grapes that hang in one of the ancient rooms. (*Vinsanto*, or holy wine, is made by hanging white wine grapes to dry for four months, then crushing the sweet "raisins" and aging the resulting wine in casks for many years.) Not a hundred yards from that room, a little below the medieval dwellings, stand gleaming, new stainless steel holding tanks—the latest in wine technology—that contain the year's harvest of Chianti Classico for one of the area's fine wine producers.

In the Chianti countryside one is always aware of the oldness of things. Life five hundred years ago was certainly no less sophisticated than it is today. Indeed, throughout the Middle Ages and the Renaissance, and after, Chianti was of great political, intellectual, artistic, and agricultural importance. Each of the region's ancient castles, abbeys, and estates can trace its history through letters, accounts, maps, and so on, in surprisingly intricate detail. At the castle of Uzzano, for example, the walls of the tasting room where olive oil and wine are sampled are hung with documents dated hundreds of years before. A letter marked "Prato, 1398" mentions the wine of Uzzano; a prize for olive oil from an American exhibition in 1876 hangs next to a photo of a seven-foot-high, prize-winning Chianina bull raised on the estate.

While visiting the Chianti region in the fall of 1982, I was fortunate to make the acquaintance of a charming and amiable American, John Meis, who lives at an old abbey in Chianti, close by the ancient towns

N. 020124

of Gaiole, Radda, and Castellina. The abbey and its estate are owned by Piero and Lorenza Stucchi-Prinetti; his family has been the proprietor since shortly after the abbey was secularized by Napoleon. The farm is managed from the beautiful twelfth-century, castle-like pile, and produces the finest of olive oils, Chianti Classico wine, *vinsanto*, and farm products typical of the area, such as an unforgettable chestnut honey.

The Badia a Coltibuono (the Abbey of the Good Harvest) is an estate of about 2200 acres, of which only a small portion is planted with olives. Another small percentage of land is devoted to grapes; the rest is pine and chestnut forest. Here, the olive trees are grown at a high elevation—1500 feet (2000 feet is the limit for olives). Most of the abbey's seven thousand trees are of the Frantoio variety, a low-yielding, cold-susceptible olive which produces one of the fruitiest, most peppery, and least fatty and acidic of the oils. Growing among the Frantoios are some heartier Leccinos and Maraiolos, and some Pendulinas, which fertilize the Frantoios. (On some of the neighboring farms, the faster-ripening Leccino olives are harvested for making a traditional salt-, orange peel-, and fennel-cured table olive.) None of the olive trees is large; they stand only twenty feet from one another without coming close to touching. And at this elevation even the oldest of the trees remains small.

Like many of the old estates, Coltibuono has adopted modern techniques whenever they can ensure a better product. Dr. Maurizio Castelli, a young enologist and agronomist, oversees the oil and wine production at the abbey. He has introduced new methods of production in the abbey's Chianti Classico wine making, but has left the ancient techniques for making *vinsanto* intact. Passionate about quality, he has left less to chance with the olive oil production than was done in the past.

Coltibuono. Avenue of lime trees in foreground leads to the most ancient part of the abbey. Campanile is XII th century.

Maurizio explains that because the trees are no longer planted on terraces, tractors can be used to till the soil. He quotes the old adage, "The olive tree is generous to the generous cultivator," and says that the soil must be turned twice a year. Although the trees can do fairly well by themselves on this arid, rocky soil, Maurizio uses nitrogen and organic fertilizers, and says that some of his neighbors have even introduced drip irrigation. The northeastern winds that seem so harsh to us are a blessing because they help stave off the olive's natural pest enemies. (That is essential because pesticides cannot be used on an oily fruit that retains chemicals and odors.)

He trains his workers particularly well in pruning. The old ways were not the best ways, and Maurizio's scientifically developed methods alleviate the previous tendency to "alternate bear," and get the most from each tree. Unlike the grape, the more the olive bears, the better the fruit.

At the abbey, as at most of the olive oil-producing farms in Tuscany, the olives are picked when they are a lovely red ripe. (Although they have a higher oil content, the riper purple and purple-black olives are much "fattier" and more acidic and are to be avoided when making fine olive oil.) The olives become red during November and December, when the days are their shortest and the air is raw. The silver leaves of the small, wind-bent trees look grey against the misty sky. The same families that just two months before had picked grapes in the autumn sun are now warmly dressed against the cold for the arduous hand-harvest. The pickers pluck the olives one by one, just as they were plucked hundreds of years ago, lest the olives be bruised or the trees damaged.

As are most of the olives in the area, the abbey's harvest is processed at a local *frantoio* (after which the variety of olive is named). The operation at Pisignano is like many of the mills in the area, including the one to which Coltibuono takes its oil. It is run by Valerio Balbieri, and about a hundred and fifty farmers depend upon him for their pressings. Balbieri's workers are paid in part in oil, as are many of the agricultural workers in Tuscany. Their tools lie atop an ingeniously constructed receptacle which catches all the oil that drips off the scoops and cups. During December and January, the small, tile-lined building is alive with festive activity twenty-four hours a day, with the exception of eight hours at Christmas and eight hours on New Year's. The farmer whose olives are being crushed and pressed watches, drinks wine, and jokes with other farmers and the mill workers while the oil is extracted from his harvest. At Pisignano, the giant, new eight-ton stones are run by machinery, not by oxen or "blind" horses as they were in the past. The crushed olives, or "mash," are layered on strawlike mats, and the old mechanical presses do their job. Then the water is spun from the oil by the small centrifuge, and the proud farmer soaks a piece of bread in his oil, tastes it, and pronounces his the best oil ever pressed.

The young, college-trained Maurizio from the abbey is no less proud of his product than are the earthier-looking farmers. The abbey's oil, once tasted, is taken back to be stored in *orci*, the enormous old earthen urns traditional in Tuscany for hundreds of years and, like the *orci* of most of the oil makers of the area, wearing the patina of oils from many years' harvests. The oil sits in cool, dark caverns, not unlike wine cellars, until it is bottled with a decrepit, antique-looking bottling machine.

But before the new oil is poured into the lovely, tall, hand-blown bottles of the abbey, it is filtered through cotton. The procedure is not wholly necessary, but the turbid appearance of the unfiltered product is far less attractive than the clear, green-gold filtered oil, and Maurizio maintains that filtration permits longer storage.

The fine, fruity, peppery oil is a staple in the Stucchi-Prinetti household. It is served daily from a cruet at the dining room table, at special dinners and everyday lunches alike. The oil is poured, along with a little of the abbey's piquant wine vinegar, over salads composed of rocket and lettuces fresh from the kitchen garden, or unstintingly into minestrones. Accompanied by the abbey's Chianti, dinner is preceded by crude crackers made at the baker's in nearby Gaiole with Tuscan oil, and followed by hard, Gaiole-made *biscotti* dunked in sweet *vinsanto*.

The Tuscan oil is one of the things that contributes to the *dolce vita* of the area that my well-traveled American friend John loves so much. Nowhere, in his experience, do people savor each moment as they do in this land of the vine and the olive tree.

Alfo — vialacea var.

Olives at the Table

I know of nothing more
appetizing on a very hot day
than to sit down in the
cool shade of a dining-room
with drawn Venetian blinds,
at a little table laid with
black olives, *saucisson d'Arles*,
some fine tomatoes, a slice
of water melon, and a pyramid
of little green figs baked in
the sun. . . . In this light air,
in this fortunate countryside,
there is no need to warm
oneself with heavy meats
or dishes of lentils. The
midi is essentially a region
of carefully prepared
little dishes.

Pampille
(pen name for Madame
Léon Daudet,
in Elizabeth David's
French Provincial Cooking)

*P*lucked from the tree, at any stage of ripeness, the olive is acrid and inedible because of the glucoside that is abundant in its flesh. It seems less promising a food than an uncooked artichoke. Man has lived up to the challenge of the olive many hundreds of times over, however, by producing uncountable variations of delectable cured olives.

CURING TECHNIQUES

How it was discovered that leaching makes the olive edible no one knows. Perhaps one day some olives fell from a tree atop a cliff onto a Grecian beach, lay exposed to the sun and salty water, were spied by a particularly hungry ancient, and an important culinary discovery was made. How ever it first happened, there are now five basic ways to leach the glucoside from olives:

♦ Oil-cured are soaked in oil from one to several months.

♦ Water-cured are soaked in water, rinsed, and soaked again for many months.

♦ Brine-cured are soaked in a salt-brine solution from one to six months.

♦ Dry-cured are cured in salt from one to several months, sometimes rubbed with oil and called oil cured.

♦ Lye-cured are soaked for a few days in a strong alkaline solution made most often with lye, but sometimes wood ash or caustic soda.

After the glucoside is vanquished, the relatively simple task of adding seasoning is all that remains. Some unusual, perhaps unique, methods for curing olives can be added to that list of techniques. In the French Midi, olives called *Fachouïlles* are cured in the sun. As would be expected, they are quite bitter; they are a rural olive and are not exported. A more famous, but nevertheless unusual, method for curing olives produces *olive schiacciate* in Calabria and Sicily. The olives are picked green, crushed, then cured in oil, and are considered a salad. These pungent olives are not available abroad. Descriptions of a variety of cured olives from olive producing countries from around the Mediterranean follow.

Spain. Most of the world's table olives come from Spain. The plain around Seville in Andalucia has the perfect climate and soil for the mass production of olives. The common varieties are the Manzanillo (a variety that was transplanted with great success to America), the fat Gordal (which is much like the Sevillano, another transplant to America), and the Picual. Well over two million metric tons of olives are produced in Spain yearly by large companies (at least one of which is American-owned), and about two-thirds of the crop is converted to the green, factory-processed olives we see most often, stuffed and packed in glass jars. While still green, they are beaten from the trees with sticks by peasant workers and farmers, given a short lye cure, then fermented in salt brine and preserved with lactic acid. All that remains of the olive is its shape and firm texture, the olivy flavor having been replaced by salt and acid.

The Spanish have developed state-of-the-art machinery: a pitter that processes 1700 olives a minute. (California olive producers brag about a 1000-pit-per-minute machine.) The Spanish also manufacture the same, reconstituted, gelatinous pimiento the Americans do for stuffing their green olives, and are working on a similar substance made with anchovies. A machine that replaces the top on the pitted olive will cover the implantation without leaving so much as a scar, making the new, anchovy-stuffed olive one of the great accomplishments of the industrial age. However, the hand labor necessary for stuffing olives with onions and nuts has yet to be replaced by machinery.

It is unfortunate that no rural, brine-cured olives are exported from Spain into the United States. There are still small olive farmers in Spain—and not only in Andalucia, but farther north, in Catalunya—who grow Arbequiña and other more oily olives and cure them for local consumption. Many Spaniards think of the flavor of cracked green farm olives when they think of Christmas, for that is the time of year when the new olives are first ready to eat.

France. It has been said that in Provence alone there are three hundred different types of olives available in village markets. The French are extraordinarily imaginative when it comes to inventing ways to process and flavor olives, although one of their most famous olives, the picholine, is named after an Italian expatriate, Picholini,

Olive gatherers in Andalucia.
(after a photo in The Classic Cooking of Spain.*)*

who is said to have introduced the use of ash curing to France. *Olives en saumure* are red-ripe, needle-pricked olives, leached in water for ten days before soaking in brine; *olives farcies* are variously stuffed olives; *olives cassées* are cracked, green olives, cured in water, then marinated in a salt brine with fennel.

One advantage the French have in producing splendid olives is a splendid climate—a colder climate that produces more flavorful, albeit smaller, olives than the hotter Mediterranean climates. The famous olives of France—the picholine, the Nyons, and the Niçoise—are tiny when compared to the plump Gordal of Spain, but have superior texture and flavor. Other French varieties are Rougeon, Redoutant, and Coucourelle.

Italy. Italy produces both small and large olives because of its wide range of climates. In the oil olive country around Tuscany, farmers use the small Leccino to make a table olive flavored with orange peel and fennel. Tiny, black, wrinkled olives are sold in the area around Rome; on the streets of Rome mild, medium-size *olive dolce* are sold in paper cones. Castellamare is famous for its black olives. From Liguria come the subtle, small ripe olives that are aged eight months in brine. One of the Gaeta olives is tiny, purple, and brine cured; another is larger, meatier, and dry cured.

Greece. Greece is the number-two table olive producer in the world and exports a much larger variety than Spain. The United States imports more types of brine-cured olives from Greece than from any other country. The large olives of Greece are picked at all stages of ripeness. Greece's most famous green olive is the Agrínion, a cracked olive the color of a green sea. The Amphissa is a purple olive, indicating it is picked almost dead ripe, and is brine cured. On the islands around Míkonos, farmers cure their large olives themselves and produce one that is similar to the Kalamata in flavor: They slit it in three places, soak it in water for two weeks, then store it in a salty brine (brine is judged salty enough when a whole raw egg floats in it) to which a little vinegar is added after eight hours.

*The olive vendor in Carcassonne.
(After a photo by R. Klein.)*

Eastern and Southern Mediterranean. The eastern and southern Mediterranean areas produce olives that never see American shores, or, it is said, are sold in the United States as Italian olives. In spite of the fact that Syrians, for example, cure olives at all stages of ripeness and produce a variety of good olives, their olives are sold almost solely at domestic marketplaces. (Local housewives store them in lightly vinegar-flavored water and float olive oil on the surface.) Of Tunisia's Ouslati and Meski olives the same can be said. Among all the African olives, only Morocco's have appeared in our deli cases. Morocco makes hundreds of different olives and they are used extensively in Moroccan cuisine, but I have seen only two here: a round, brine-cured olive (see glossary, page 62), and a meaty, dry-cured olive, perhaps the best of its kind available.

IMPORTED OLIVES

The "ethnic" olives that are shipped to the United States come either in plastic, five-gallon containers or in large, beautifully decorated cans, packed in their cloudy mother brine. The trip takes about six weeks from farm (one would hope) to retail outlet. Because olives are "alive" (they continue to ferment in their brine), shippers have sustained some losses when gases cause cans to explode. Inventive packers have designed cans with valves to allow the gases to escape. Others pasteurize the olives, stopping the fermentation process but destroying some of the subtle flavors of the olives. The longer an olive is allowed to ferment in its own brine, the less bitter and more intricate its flavors will become. Some olives age eight months, others as little as two months.

AN OLIVE GLOSSARY

The following index of olives represents most of the general types of olives now available in specialty shops around the country. It is by no means exhaustive, nor does it touch on the marinades and herb flavorings prepared at individual stores. Furthermore, a greater variety of olives is being imported all the time. This index simply gives a sampling of the many curing procedures, sizes, and shapes American cooks now can count on as ammunition in their culinary arsenal.*

Accurate names for the olives have yet to be devised. Most are sold by the name of the area or town of origin (e.g., Gaeta and Niçoise); several are named for the variety of olive from which they are made (e.g., picholine and Salona); still others are sold according to the method of curing (e.g., Greek- and Sicilian-styles). Confusion arises when more than one olive comes from one place, as is possible with all olives, no matter how small the place of origin; when one variety is cured many different ways; when one person's Greek-style olive is another person's dry cured; or when olives are deceptively labeled (many olives said to have come from Gaeta, for example, are actually from less prestigious areas of the Mediterranean). That is why it is a good idea to sample olives before buying. Store owners are usually very obliging, and tasting will save you surprises when you get home. You can determine, too, whether an olive has been around too long and has become soft.

It is advisable to bring your own lidded jar with you so that you can store the olives in the mother brine. If you do purchase a carton of unbrined olives and plan to keep them for more than a few days, it would be best to make a brine or add olive oil to the olives. Your olives can be kept out of the refrigerator in a cool place for a number of weeks. In fact, some importers claim that, properly kept in a salty brine, olives can last upwards of two years.

If certain olives are too strong for a dish you are planning to make with them, soak them in plain water for a few days or dunk them in boiling water for about 10 minutes. The texture of green olives especially will not suffer from the brief boiling, although the color will change to brown-green. To bitter, dry-cured olives add oil, garlic, and herbs; they will help balance the strong olive flavor.

---

*The olive drawings are to scale.

ITALY

Liguria

(Available currently only on the East Coast.) Black to brown-black. Salt-brine cured. Cured for eight months. Tasty, alive. Sometimes packed with stems. Very slightly acidic.

Lugano

Very dark purple-black. Salt-brine cured. Rather salty. Sometimes packed with a few small olive leaves. A popular olive at tastings.

Ponentine

Purple-black. Salt-brine cured, then packed with vinegar. A mild olive.

Gaeta

Black and wrinkled. Dry-salt cured, then rubbed with oil. Surprisingly mild. Often packed with rosemary and other herbs. Other olives bearing this name are brine cured.

CALIFORNIA

Sicilian-style

Medium green. Salt-brine cured and preserved with lactic acid, either added or produced by the olives themselves. Very crisp. Made from California's largest variety, the Sevillano. One variation is a cracked green that can be cured a shorter length of time.

Dry-cured

Black and wrinkled. Dry-salt cured and rubbed with olive oil. Flavorful and meaty.

Greek-style

Black-purple. Salt-brine cured and packed with vinegar. Size can vary. Fairly firm fleshed.

Black-ripe pitted

Shiny, unvarying black. Lye-cured, with color produced by oxygenation and fixed with ferrous gluconate. Harvested green. Canned in a mild salt brine. Ten sizes in the United States. Now also produced in Spain.

GREECE

Naphlion (or various other spellings)
Dark green, cracked. Salt-brine cured, then packed with olive oil. A crisp olive with a pleasant, youthful bite.

Salona
Brown to brownish-purple. Salt-brine cured. Soft to mushy texture, salty but pleasing flavor. Has a small pit.

Kalamata (or Calamata)
Black-purple. Slit, then brine cured and packed with vinegar. Size can vary. A lovely, almond-shaped standby, appreciated by many.

Royal (or Royal Victoria)
Red, dark brown, light brown from same batch. Slit, then salt-brine cured and packed with vinegar and olive oil. Similar in flavor to the Kalamata. Size varies considerably. Often shipped in bulk and packed in bottles by the importer.

MOROCCO

Morocco

(The only name I know for this olive, although there are hundreds of different kinds of olives in Morocco.) Black-red to black. Salt-brine cured. Packed with twigs and leaves. Fairly firm-fleshed for such a large ripe olive. This pretty olive is almost perfectly round. Popular at olive tastings.

Dry-cured

Black and wrinkled. Dry-salt cured. Meaty. Rather bitter. Nicest when rubbed with oil and mixed with garlic and spices by merchant.

FRANCE

Nyons

Black, with green tint. Dry-salt cured, then rubbed with oil. Rather bitter. There are brine-cured olives from Nyons as well.

Picholine

Medium green and smooth. Salt-brine cured. Subtle, lightly salty. Sometimes packed with citric acid as a preservative when bottled in the United States. (Named after the variety of olive.)

Niçoise

Brown to brown-green-black. Salt-brine cured. A tasty, quite small olive with a high pit-to-meat ratio. Often packed with herbs, stems intact.

PERU, BRAZIL, CHILE

Alfonso

Blue, purple, black-brown. Salt-brine cured, then packed with vinegar. Olive oil is often added by the importer when it is bottled.

SPAIN

Spanish-style

Bright green. Variously stuffed or unpitted. Lightly lye-cured, then packed in salt and lactic acid brine. Various sizes. Also produced in California.

PITFALLS

There are many recipes that call for pitted, brine-cured olives. Because such an article cannot be purchased at any price, I set out to find a suitable implement so I could produce my own pitted olives in no time. I made a complete tour of the stores in my area. All I could find was something called a cherry pitter, and a new French device (probably of medieval design) I found in New York labeled a *dénoyauteur*.

The cherry pitter worked perfectly well on certain ripe olives that were not too soft, not too hard, had a loose pit, and were of a precise, moderate size. When I tried pitting a picholine with it, I poked a hole in my thumb big enough to stuff a pimiento into. The cherry pitter was worse than useless.

The *dénoyauteur*, on the other hand, worked well on all the smaller, ripe, brine-cured olives. If you come across one, and are in the habit of making Provençal tapenades, it will be worth the investment. I have read of a *chasse noyaux*, which one can buy in France and is reputed to be adept not only at cracking but at pitting olives. From the general sound of the words *chasse noyaux*, my suspicion is that the instrument is likely to crush any olive it gets its teeth into.

Still, cutting the meat from the pit with a sharp paring knife is not so tedious a task. With certain cracked and slit olives, the pits can be removed easily without the aid of a knife, and the olives remain in one piece—sort of.

RECIPES

ABOUT THE INGREDIENTS

All the recipes compiled here have either olives or olive oil as a major ingredient. The recipes are Mediterranean in flavor, except for a few that are distinctly Mexican or Californian. Here is a list of some of the special ingredients found in the recipes and information concerning their use.

Olives

In the recipes that use olives, I designate the appropriate variety for the dish. If the variety is not available to you, refer to the Olive Glossary (pages 58 to 64) for olives that are similarly made and that can be substituted.

Olive Oil

Recipes using olive oil call for either "fine olive oil" or "olive oil."

"Fine olive oil" is to be used as a condiment; its use in cooking is an extravagance, since the flavors of the expensive oil are altered at cooking temperatures. Fine olive oil should be your favorite oil, the oil with which you dress a salad or finish a chop, or serve at the table from a cruet. I have broken this category down further for use in my home: I have a very peppery, fruity oil I serve on rich or highly flavored dishes, and a slightly milder oil for salads made with delicate lettuces.

"Olive oil" refers to less expensive oil that need not have the attributes of a fine oil because it is used in cooking and would be altered by heating. (See pages 39 to 41, for information on purchasing oil.)

Herbs

Most of the herbs used in the recipes are specified "dried" because fresh herbs are often not available. As a general rule, fresh herbs may be substituted for dried herbs by more or less doubling the amount. Certain recipes call for fresh herbs and, with the exception of rosemary, cannot be substituted for. (Basil and tarragon, for example, have a completely different character dried. Oregano, on the other hand, is quite pleasant in its dry form—even preferable to the bitter fresh herb.) Fortunately, herbs are easy to grow in window-sill boxes, on decks, and in herb gardens. Parsley, lemon zest (and juice), cilantro, and garlic are available fresh all year and must never be substituted for. Garlic is fresher and sweeter at certain times of the year than at others. Old garlic is bitter, and often has begun to sprout. California bay leaves are coarser, more overpowering in flavor than imported ones.

Anchovies

Canned anchovies, packed in oil, are quite salty and should be rinsed before they are used. Some delicatessens and specialty shops carry *anchois salé*—salted anchovies that come in bulk and are sold by the ounce. They too must be rinsed, but are more delicate than and much preferable to canned anchovies. If you do not use all you buy, store the rest in olive oil in the refrigerator.

Feta Cheese

Many of the recipes call for feta cheese—a fairly soft, sheep's milk cheese beloved by the Greeks. The United States imports feta not only from Greece but also from Rumania, Bulgaria, Israel, Hungary, Corsica, and Denmark. There is a domestic version as well. The Bulgarian feta is often the creamiest, but it is wise to ask your cheese merchant which variety he likes the best and why, then make up your own mind. Prices can vary considerably.

Tomatoes

Although fresh tomatoes are available throughout the year, good tomatoes can be purchased only a few months during the summer and
fall. Canned tomatoes are far preferable to the mealy, pink, flavorless
tomatoes we get during the rest of the year. Again, tomatoes are easy
to grow and there is nothing like a vine-ripened tomato fresh from
the garden.

Pumate

A few recipes use sun-dried tomatoes, or *pumate*, available at specialty food stores. Some are better than others, as you will see when
you compare. *Pumate* are quite expensive, but a little goes a long
way. The delicious, sweet fruit comes packed in olive oil and will
keep for weeks in your refrigerator.

Butter

This *is* an olive oil book, so there aren't many recipes that call for
butter. When they do, use unsalted (or "sweet") butter. By the time
salted butter reaches the consumer, more often than not it has been
around for quite a while. Unsalted butter can be purchased by the
pound and frozen in small quantities for everyday use.

Salads

The proper proportion of oil to vinegar for salads is as common a topic
for discussion as the proper proportion of gin to vermouth for mar-
tinis. The exact proportions will depend on the tartness of the vinegar,
the fruitiness of the oil, the bitterness of the greens, not to mention
the tastes of the cook; but, that being said, I believe there is nothing
quite as honest or so delicious as a well-made, Tuscan-style salad, for
which oil is measured in tablespoons and vinegar is measured in teaspoons—if any measuring is done at all. That perfect salad will
consist of the freshest garden greens, the cook's favorite fine olive oil,
and a smattering of first-rate wine vinegar to pour over them, and a
little salt and freshly ground pepper.

Breads

Many of the recipes in this book are best if accompanied by a Tuscan-style bread, that is, a light, very crusty bread made with no salt and no fat, not even olive oil. Such bread is difficult to find or to make, so substitutions are in order. The best is probably a sweet Italian or French bread made by a good bakery.

Baguettes are the long, variously flavored (sweet and sourdough in this country) crusty loaves of white bread modeled after the common loaf of France. Unfortunately, they may not be readily available to you. The baguette's shape makes it essential in certain recipes, while in others, any crusty French- or Italian-style white bread will substitute.

LITTLE RECIPES

Andalusian "French toast." Fry bread slices that have been dipped in beaten egg in ½-inch of olive oil. When the toasts have browned on both sides, remove them from the pan and add a few tablespoons of honey to the pan, adding more olive oil if none is left. Rub the toasts in the honey and oil and serve.

Pinzimonio. Fine olive oil, coarse salt, and freshly ground black pepper compose a simple sauce into which Tuscans dip vegetables, including raw young artichoke leaves, crunchy fennel bulb, and un-cooked green beans.

Belgian endive salad. Soak 3 Belgian endives in ice water for half an hour. Shake the water out. Cut them crosswise into very thin pieces and, in a salad bowl, toss with 2 tablespoons fine olive oil, about 10 Kalamata or other vinegar-cured olives (pitted and finely chopped), 1 teaspoon red wine vinegar, salt, freshly ground pepper, and 2 anchovy fillets (rinsed and finely chopped). Serves 3.

Oven-fried potatoes. Pour 1 cup olive oil into a 12- by 18-inch baking pan. Slice 4 or 5 Russet potatoes into ⅓-inch-thick rounds. Rub the slices with olive oil on both sides and place them in the pan, sprinkling them with salt. Bake half an hour in a moderate oven; turn with a spatula, and bake until crispy brown (another half an hour or so). They make a nice accompaniment for roasts or highly seasoned meat.

Pasta with olive oil and garlic. In 2 tablespoons of olive oil, sauté 10 cloves of chopped garlic until they are crisp and light brown. In a large pot of salted, boiling water, cook 1 pound of dried pasta until *al dente*. Drain well in a colander. Return to the pot and toss with the garlic pieces and ⅓ to ½ cup fine olive oil; salt to taste and add a good sprinkling of freshly ground black pepper. Serves 4 for a hearty first course.

Broccoli sautéed in olive oil. Parboil broccoli flowerets. Drain. Then sauté them quickly in plenty of olive oil with crushed garlic. Salt and pepper to taste. Makes a simple accompaniment for just about anything, made with a vegetable that never seems to go out of season.

Fruit cake. It is wonderful made with olive oil. A good combination is half butter and half olive oil.

Marinated red bell peppers. Scorch the skins of red bell peppers under the broiler or on an open gas flame. Rub off the skins under cold running water. Sauté the peppers in olive oil until they are limp with plenty of crushed garlic. Layer in a bowl and season with salt, freshly ground black pepper, and a drizzle of fine olive oil. (You may cut the oil with a few drops of lemon juice.) Serve at room temperature as an accompaniment to a roast or plain grilled fish or chops.

Crostini. Crostini are toasts covered with spreads or toppings and served as hors d'oeuvres or as a first course. To make crostini, use a fine-textured white bread—even an egg bread—thinly sliced and toasted, then fried in olive oil, crushed garlic, and salt. Serve with a variety of toppings such as the tapenades on pages 88 and 89. Other toppings include:

— thinly sliced Parmesan cheese (sliced with a vegetable peeler) with fine olive oil drizzled over it;

— *olivade*, made with ½ pound whole milk ricotta, ½ pound finely chopped brine-cured black olives (such as Niçoise or Ponentine), 1 tablespoon Cognac, and 1 tablespoon finely chopped *pumate*;

— a mild cheese melted on the toast under the broiler and topped with mushrooms that have been sliced and marinated in a vinaigrette;

— chopped chicken livers that have been sautéed in olive oil with onions, then mashed in a bowl with fine olive oil and salt and freshly ground black pepper to taste.

Home-Cured
Olives

CURING YOUR OWN

*A*ll varieties of fresh, uncured olives, at all stages of ripeness, are inedible. Taste one and you will realize what a miracle the cured olive is. Curing leaches the bitter glucosides from the fruit's flesh. Lye treatments (old farming communities around the Mediterranean use ash) do the job most thoroughly, producing in just a few days fruit so bland that it must be doctored with salt, and sometimes herbs and other seasonings, for flavor. Water or brine treatments produce the tangy olives of the type most frequently imported in bulk—whether they be cracked greens, purple Kalamatas, or tiny Niçoises—and take anywhere from ten days (in the case of small cracked or needle-pricked olives) to three months (in the case of large, whole black olives) to leach out the bitterness. Some olives are aged in brine as long as eight months.

Making your own home-cured olives offers the same satisfaction as putting up any fruit or vegetable. Unfortunately, it is difficult to obtain fresh olives. In grocery stores in most parts of the United States—even in California, where 99 percent of the nation's commercial olive acreage is found—fresh olives, at any stage of ripeness, are a rarity. In urban markets, I have yet to see more than one variety offered at one time. In northern California, Mission olives (a fruity, oily but small variety) have found their way into certain vegetable markets; in other parts of the state, Manzanillos (the state's most commonly grown olive and one that is somewhat larger than the Mission) might be more commonly seen because they are the most widely grown variety in the Central Valley. A small quantity of Barounis is shipped to the East Coast for home pickling, where traditionalists who still pickle their own olives are numerous enough to warrant shipment. If you do have a choice, however, use the smaller olive— the huge Sevillanc, the country's largest variety of olive, has a low oil content and is less tasty than even the slightly smaller Ascolano. If you know of a store that sells fresh olives, you are in luck—but beware of damaged olives, especially ripe ones. Often they are culls or have been so bruised in handling that they will produce a sorry pickled olive.

If you are in an olive-producing area, or have a friend or neighbor with a producing tree, or own one yourself, there is no style of cured olive you won't be able to produce, size being the only limitation. You will be able to pick olives at all stages of ripeness, carefully, one by one, and pickle them immediately. Especially when the olives are black (as in making the salt-cured type) and red (as in the Greek-style recipe), they are best just picked. Pickle the olives as soon after they are picked as possible; bruises start to show up soon after picking.

On the subject of making your own olive oil: Home economists are never at a loss for projects that take a long time, are messy, and produce ghastly results. Such projects, like the one for making olive oil at home, are on a par with a Girl Scout project I did as a child, making centerpieces out of pine cones, pipe cleaners, and styrofoam balls. You can make your own olive oil with an automobile jack and boards rigged up as an improvised press. Good advice: Buy your olive oil instead. There is a reason good olive oil costs as much as it does, and it goes beyond labor costs.

❧ ANZONINI'S WATER-CURED GREEN OLIVES ❧

This pungent recipe was given to me by a Spanish gypsy; it is almost identical to the method for making the green olives of Provence called *olives cassées*.

5 pounds green mature olives
1½ quarts water
3 tablespoons salt
2 lemons, cut into ½-inch cubes
2 tablespoons dried oregano
2 cups white wine vinegar
6 cloves garlic, peeled and halved
2 tablespoons cumin seeds, crushed in a mortar
Olive oil

Crack the flesh of the olives with a rolling pin, or by hitting each one individually with a hammer. Rinse with cold water. Place them in a stoneware, earthenware, glass, or porcelain jar and cover with cold water. Weight them with a piece of wood or a plastic bag filled with water (to keep the olives submerged) and keep them in a dark, cool place for ten days, changing the water every day.

Boil the water and dissolve the salt in it. Empty the liquid from the jar in which the olives have been soaking; rinse the olives in cold water and cover the olives with the salt brine. Mix in the lemons, oregano, vinegar, garlic, and cumin. Float enough olive oil on top to cover the surface. Store in a cool place at least two weeks. To make a more interesting mixture, add a few store-bought Kalamata olives. Store in a cool, dark place. The olives keep quite well for at least two months.

❈ SALT-CURED RIPE OLIVES ❈

These flavorful, if bitter, shriveled dry-cured olives—sometimes called oil-cured—will not keep nearly as well as brine-cured olives. Because of that and the fact that they are so pungent and not to everybody's liking, you might want to make only a small quantity of them. Use olives that are black or almost black. Mission olives are the best because of their high oil content and small size. Extra-large olives, such as the Sevillano, become soft.

Cover the bottom of a thick cardboard or wooden box with burlap or cheesecloth. In the box, mix together equal weights of non-iodized salt and olives. Spread out evenly; then pour a layer of non-iodized salt over the olives so that nearly all of them are covered, using an additional pound or so of salt. Place the box outdoors in the shade or in a basement so any liquid that oozes from it will not stain a floor or decking.

Stir the salt-covered olives well with a wooden spoon once a week for four weeks, or until the olives are cured. They should be slightly bitter.

Remove the olives from the salt by hand (unfortunately, I have found no better method). Dip the olives in a large pot of rapidly boiling water for a few seconds; then drain in a colander and refresh with cold tap water. After spreading them out on paper towels, let them dry for a few hours or overnight. Those olives you wish to eat within a few days should be coated with fruity olive oil (rub them with your fingers to distribute the oil), mixed with your favorite herbs, and kept in the refrigerator in a tightly capped jar. The remainder of the olives should be mixed at a ratio of two parts olives to one part non-iodized salt by weight and kept in a cool place or refrigerated. They do not keep more than a month.

❄ *GREEK-STYLE RIPE OLIVES* ❄

*F*or this recipe, choose olives that are red to dark red. Slash each olive deeply on one side using a very sharp knife to reduce bruising. Place olives in a large stoneware, earthenware, glass, or porcelain container. Make a solution of 4 tablespoons salt dissolved in 1 quart water, and pour enough over the olives to cover; then weight the olives with a piece of wood or a plastic bag filled with water so that all of them are completely submerged. Store in a cool place, changing the solution once a week for three weeks. If a scum forms on the surface during that time, disregard it until it is time to change the brine; then rinse the olives with fresh water before covering with brine again. The scum is harmless.

At the end of three weeks, taste one of the largest olives. If it is only slightly bitter (these olives should be left with a bit of a tang), pour off the brine and rinse the olives.* Then marinate them with the proper amount of liquid to cover in a marinade made according to these proportions:

1½ cups white wine vinegar
1 tablespoon salt dissolved in 2 cups water
½ teaspoon dried oregano
3 lemon wedges
2 cloves garlic
Olive oil

Float enough olive oil to form a ¼-inch layer on top of the marinating olives. The olives will be ready to eat after sitting in the marinade for just a few days. Store, still in the marinade, in a cool pantry, or in the refrigerator. If kept too long, the lemon and vinegar flavors will predominate—so eat these within a month after they are ready.

---

*If the olives are too bitter to be put in the marinade, rebrine and soak for another week; then rinse and marinate.

❧ *LYE-CURED GREEN OLIVES* ❧

*O*n the street corners of Rome these sweet olives (*olive dolce*) are sold by the handful for next to nothing, wrapped in paper cones. They last about a city block. They are a lovely, bright green color, and are buttery in flavor. Marinate them according to the recipe on page 81 if you prefer a spicier flavor with a touch of vinegar.

Use olives that are mature but still green. Purchase lye in the "cleanser" section of your grocery store.*

Rinse the olives with water and place them in large glass or porcelain jars; then determine how much lye solution you need to cover the amount of olives you have. Add a solution that has been mixed at the ratio of 1 quart water (at 65° to 70°F) to 1 tablespoon lye. Soak 12 hours.

Drain olives; then soak 12 more hours in fresh lye solution. Drain and rinse. Cut into the largest olive; if the lye has reached the pit, the lye cure is complete. Rinse again and soak in cold water. (Usually two lye baths are enough for the small Mission olives seen in specialty produce stores.) If one more bath is necessary, soak in fresh lye solution for 12 more hours; then drain and rinse with cold water.

Soak the olives in fresh, cold water, changing the water three (or more) times a day for the next three days. At the end of three days, taste an olive to make sure that there is no trace of lye flavor remaining.

Next, soak the olives for at least one day in a brine solution mixed at the ratio of 6 tablespoons salt to 1 gallon water. The olives are now ready for eating. Store the rest in the brine solution in a cool, dark place, preferably the refrigerator, or marinate (see recipe, page 81) and store in the refrigerator. Use within two months.

---

*WARNING: Lye can cause serious burns. Keep lemon or vinegar handy to neutralize any lye that splashes onto the skin. If lye gets into your eyes, bathe them with running water and call your doctor. If lye is swallowed, call your doctor, drink milk or egg white, and do not induce vomiting.

❧ MARINATED BLACK OLIVES ❧

Some would rather not bother with the canned California olive because it is so bland. Others appreciate it for its firm flesh, which is firm because it was picked green, whether it winds up green or black. The following two recipes give ways to add flavor to the unobtrusive, commercially prepared fruit, or dimensions to the more flamboyant imports. Using unpitted olives helps the aesthetic.

8 ounces California "black-ripe" or any cured black olive
½ teaspoon dried tarragon
½ teaspoon dried thyme
1 tablespoon coriander seeds, crushed in a mortar
2 bay leaves, broken
6 large pieces orange zest
¼ cup white wine vinegar
2 tablespoons salt dissolved in 2 cups water
Olive oil

In a jar just large enough to contain olives and brine, alternate the olives with tarragon, thyme, coriander, bay leaves, and orange zest so that the spices are well distributed among the olives. Add the vinegar; then cover with the brine. Float enough olive oil on top to cover the surface of the liquid. Then cover the jar and refrigerate for at least one week. Let the olives warm to room temperature before serving, so that the olive oil coats the olives as you remove them from the jar. Use within a month for best flavor.

MARINATED GREEN OR
❧ "GREEN-RIPE" OLIVES ❧

*U*se canned "green-ripe" olives or the sweet olives described on page 79 with the following marinade. The California "green-ripe" olive is not a true ripe olive, but its curing and cooking make it a pale green-brown instead of bright green.

> *8 ounces olives*
> *1 large clove garlic, halved*
> *½ teaspoon salt*
> *1 dried hot red pepper*
> *½ teaspoon dried basil*
> *2 bay leaves, halved*
> *¼ cup white wine vinegar*
> *Olive oil*

Retaining some of the olive brine, put the olives in a jar small enough so that they fill it almost completely, interspersing the garlic, salt, pepper, basil, and bay leaves among them. Add the vinegar, and a mixture of half reserved brine and half water to almost fill the jar. Float a little olive oil on top. Cover tightly and refrigerate about one week. Let the olives warm to room temperature before serving, so that the olive oil coats the olives as you remove them from the jar. Use within a month for best flavor.

Hors D'oeuvres

❧ TUNA AND OLIVE ANTIPASTO ❧

For years I have purchased this antipasto from my neighborhood delicatessen, a shop owned by Italians from Genoa. It is almost a pickle, and is not meant to be eaten alone. It is a fine accompaniment on a picnic, and can be used as an antipasto with cold meats, pasta, and squid salad, and Italian bread.

2 small, unpeeled boiling potatoes, cubed
5 tablespoons olive oil
1 yellow onion, thinly sliced
1 clove garlic, minced
6 tomatoes, peeled, seeded, and chopped
2 bay leaves
1/2 teaspoon dried oregano
2 carrots, peeled and sliced thick
3 celery stalks, sliced thick
6-ounce can tomato paste
2 tablespoons capers
4 tablespoons white wine vinegar
Hot red pepper flakes
6 1/2-ounce can water-packed tuna, drained
6 anchovy fillets, rinsed and torn into pieces
1/4 to 1/3 pound shiny black olives, such as Gaetas or Luganos
Salt

Steam the potatoes until just done. While they are steaming, heat 2 tablespoons of the olive oil in a saucepan. Add the onion and garlic, and sauté until soft. Next add the tomatoes and their liquid, bay leaves, oregano, and carrots; cover and simmer 5 minutes before adding the celery. Then simmer, covered, 5 minutes more; remove from the heat and add the tomato paste, capers, vinegar, a good sprinkling of the hot pepper flakes, tuna, anchovies, olives, potatoes, and the remaining 3 tablespoons olive oil, stirring gently until well mixed. Salt to taste. Refrigerate and serve cool.

Serves 8.

❧ *FETA AND OLIVE PLATE* ❧

*I*f the olive is the fruit of the Mediterranean, surely garlic is the vegetable. One rarely finds the one flavor without the other. This pungent hors d'oeuvre plate was served to me by an aficionado of garlic—a man with passionate tastes in food, my publisher.

On a large platter, arrange a layer of bite-size cubes of creamy feta. Mix crushed garlic with olive oil and sprinkle over the cheese. Don't skimp. Sprinkle dried herbs such as oregano or thyme over the cheese, as well as some finely chopped fresh parsley. Surround the cheese with Kalamata olives that have been marinated in oil, garlic, and parsley. Serve with the freshest country-style bread, and use the bread to soak up the oil.

❧ *MOZZARELLA CHEESE PLATE* ❧

*B*ecause fresh, whole-milk mozzarella is so delicate, it is a perfect vehicle for showing off your best olive oil—the olive oil with the most finesse. Fine mozzarella has begun to replace the rubbery, skim-milk variety in good delicatessens and cheese stores.

Using the best mozzarella available, cut the cheese into the thinnest slices possible. (Do it with a wire cheese cutter when the cheese is cold.) Pour some delicate olive oil onto a serving platter. Rub both sides of the cheese slices in the oil and arrange them artfully on the platter. Serve surrounded by slices of fresh, red-ripe tomatoes, with still more olive oil sprinkled over all, and accompanied by a good bread.

✿ MUSHROOMS STEWED IN OLIVE OIL ✿

This simple hors d'oeuvre nicely complements a tray of cheese before dinner. The tomatoes and mushrooms can also add color and
texture as a garnish on a monotonous plate. Use plenty of cracked
black pepper; add just a few drops of lemon juice to cut down the
oiliness.

½ cup olive oil
1 tablespoon coriander seeds, crushed
1 pound fresh mushrooms, any cultivated or commercial variety
3 bay leaves, broken in half
10 garden-fresh cherry tomatoes
Fresh thyme (or oregano or marjoram)
Salt
Freshly ground black pepper
Fresh lemon juice

In olive oil, cook the coriander seeds for 2 minutes. Add the mushrooms and bay leaves and cook over a low heat. When the mushrooms
have stopped absorbing oil, they are done. At that moment, add the
tomatoes and a few sprigs of thyme and cook about 1 minute or so—
not long enough to break the skins of the tomatoes. Sprinkle with salt.
Add plenty of pepper and a few drops of lemon juice. Allow to cool to
room temperature before serving.

Serves 6 as an hors d'oeuvre or garnish.

❀ *FRIED CHÈVRE HORS D'OEUVRE* ❀

*W*arm chèvre, or goat cheese, is being served more and more, and it's wonderful despite its stylishness. Don't be deterred by columnists who have contempt for food chic. I'll always remember when I first had warm chèvre (*chèvre chaud*—I thought I'd ordered the Chef's Show). It was presented as a first course in an elegant restaurant in Paris, served on a bed of bitter greens dressed with a vinaigrette. It is wonderful that way, or as the warming hors d'oeuvre described here.

½ log chèvre, without cinders, at refrigerator temperature
1 egg
1 tablespoon milk
5 tablespoons bread crumbs
½ tablespoon dried thyme
4 tablespoons olive oil
1 clove garlic, crushed
Baguette

Cut the chèvre into ⅓-inch rounds, about 8 per half log. Beat the egg and milk with a fork until blended. Dip both sides of the chèvre slices in the egg mixture; dredge in the bread crumbs mixed with thyme. In a heavy skillet, heat the olive oil and garlic until the olive oil reaches the vapor stage (never let it smoke) and add 4 chèvre slices. Cook 1 minute per side. Turn with a spatula, being careful not to disturb the crust. Serve the cheese hot on a fresh baguette sliced lengthwise. If necessary, add more oil to the pan for cooking the next 4 slices. Drizzle the remaining oil over the bread and cheese.

Makes 8 slices.

✻ STUFFED EGGS ✻

Devilled eggs are comforting; they make you think of simpler times, and your green lunchbox. The following recipe for stuffed eggs would, no doubt, have little appeal for children—they are for a more mature palate.

10 cloves garlic, peeled
2 anchovy fillets, rinsed
5 hard-boiled eggs
2 teaspoons capers, crushed
2½ tablespoons fine olive oil
1 tablespoon white wine vinegar
¼ cup Sicilian-style green olives, pitted and very finely chopped
1 heaping tablespoon finely chopped cilantro
Large pinch cayenne pepper
Purple onion, sliced, or roasted red peppers

Boil the garlic cloves in water for 10 minutes. Remove and let cool. In a small mixing bowl, mash the garlic cloves and anchovy fillets with a fork. Halve the eggs and scoop the yolks into the anchovy mixture. Mash thoroughly. Mix in the crushed capers. Slowly add the olive oil while continuing to stir with a fork. Stir in the wine vinegar, olives, cilantro, and cayenne pepper. Mix well and fill the egg halves.

Garnish with bits of thinly sliced purple onion or pieces of roasted red peppers.

TAPENADE

Although its origins are Provençal, tapenade, or black olive dip, has been made in many places in many ways. It has so many uses that the variations are justified—it can be spread on crackers, toast, or bread; used between layers in pastry or filo dough; or indulged in as a dip for vegetables or what have you. Here are three diverse recipes.

❦ SIMPLE TAPENADE ❦

2-ounce can of anchovy fillets, rinsed
½ cup California pitted "black-ripe" olives
¼ cup capers
¼ cup olive oil
1 tablespoon Dijon mustard

In a blender or food processor, blend the anchovies, olives, capers, olive oil, and mustard. If the tapenade is too thick, drizzle more olive oil into the mixture as you continue to blend for a few seconds more.

Serve with lightly parboiled or raw vegetables such as fennel bulbs, scallions, cauliflower, carrots, or broccoli, or with cooked artichoke leaves. Or try it on unsalted bread.

Serves 6 as an hors d'oeuvre.

❦ CALIFORNIA TAPENADE ❦

6 ounces California pitted "black-ripe" olives
2 tablespoons capers
4 anchovy fillets, rinsed and chopped
3 tablespoons olive oil
2 large cloves garlic, crushed
3 teaspoons commercial hot salsa made with onions and tomatoes
½ teaspoon dried oregano
2 teaspoons lime juice
3 tablespoons finely chopped yellow onion
Lettuce, red leaf or curly endive

8 ounces sour cream
1 lime, cut into wedges
Tortilla chips, unsalted

In a blender or food processor, blend the olives, capers, anchovies, olive oil, garlic, salsa, oregano, and lime juice. Mix in the finely chopped onion. To serve, line a shallow bowl with lettuce, add the tapenade in a mound, and top with sour cream. Squeeze lime juice over all and garnish with lime wedges. Serve with the chips.

Serves 6 as an hors d'oeuvre.

❀ TUNA TAPENADE ❀

"**W**hat do you guys think of this tapenade?" I ventured. "I don't usually like anchovies," said one friend, "but this is irresistible." "It's like a sardine sandwich in a drum—industrial-strength sardines," said one wise guy. This recipe is quite good, and makes a large quantity. If you make it ahead of time, stir before serving; the oil tends to separate. Serve with unsalted crackers, toasts fried in olive oil, or fresh vegetables.

¹/₄ cup capers
³/₄ tin anchovies, rinsed and chopped
1 cup black olives, such as Kalamatas, pitted
3 large cloves garlic, crushed
Freshly ground black pepper
6¹/₂-ounce can water-packed tuna, drained
3 tablespoons brandy
¹/₄ cup olive oil
1 hard-boiled egg, grated

In a blender or food processor, blend the capers, anchovies, olives, garlic, pepper to taste, tuna, and brandy. With the motor still running, slowly add the olive oil until all the oil has been added and the tapenade is smooth. It should have the consistency of mayonnaise. Garnish with hard-boiled egg.

Serves 10 to 12 as an hors d'oeuvre.

❧ *MARINATED CHÈVRE* ❧

*T*his simple marinated chèvre (goat cheese) makes a delicious Provençal hors d'oeuvre—and, while marinating, is as decorative as a vase of flowers in the kitchen. It is important to have a crusty bread or baguette on which to serve it, one that will hold together when saturated with olive oil. Use small, round chèvres, such as a Lezay or those made by Laura Chenel's California Chèvre, or slice a Montrachet log (without cinders) into ⅓-pound pieces and press the ends to prevent any disintegration.

> *5-ounce piece(s) of chèvre**
> *Fine olive oil*
> *Garlic cloves (6 per each 5 ounces of cheese)*
> *Sprigs of fresh thyme*
> *Niçoise olives (12 per 5 ounces of cheese)*

Place the cheese in a jar or glass that is only slightly larger in diameter than your cheese. Pour in enough olive oil to cover the cheese. Add about 6 peeled and halved garlic cloves, a few sprigs of fresh herbs, and olives. Allow to marinate in a cool place at least two days.

Serve the marinated chèvre with a baguette or sourdough bread. Fill a small bowl with some of the marinating oil so that guests can spoon the oil over the bread. Any leftover oil can be used to dress the next day's salad or to sauté vegetables.

---

*A 5-ounce chèvre serves four as an hors d'oeuvre.

❧ *BAGNA CAUDA* ❧

*I*n Piedmontese dialect, this means "hot bath" and is made there with cream rather than with olive oil. (The cooking of northern Italy uses much more butter and cream than that of central and southern Italy.) Bagna cauda is served warm over a flame with raw or parboiled vegetables and bread sticks for dipping. Pungent and garlicky, it is not a dish for the timorous. Be sure to serve it before a meal it will not stifle.

4 cloves garlic, peeled
12 anchovy fillets, rinsed
3 slices dry toast, crusts removed
2 tablespoons cream
1 cup fruity olive oil
Freshly ground black pepper
2 tablespoons unsalted butter

In a wooden bowl, crush the garlic and the anchovies with a pestle until they form a paste. Soak the toast in the cream, crumble it into fine crumbs, and add it to the paste, continuing to mash with the pestle. Stirring with a fork or wire whisk, add the olive oil, little by little. Add a generous grinding of black pepper.

In a small saucepan, melt the butter and stir in the anchovy sauce. Heat just until warm and serve in a chafing dish or fondue pot over a flame with grissini (breadsticks) and any of the following vegetables, parboiled or raw according to your taste:

broccoli, scallions,
cauliflower, fennel bulb,
romaine lettuce, carrots,
mushrooms, green or red bell peppers,
asparagus, endive,
turnips, radishes

Serves 8.

Breads,
Pizzas,
and
Sandwiches

❧ *PATAFLA* ❧
Tomato and Olive Sandwich

This is one of the many rural recipes—like tapenades and country salads—that seem to have been invented independently all over the Mediterranean but are basically the same. This patafla, a kind of sandwich, differs from *pan bagna* only in that it has no anchovies.

1 baguette, the crustier the better
Fine olive oil
3 ripe tomatoes, peeled, seeded, and chopped
½ purple onion, minced
½ pound mixed brine-cured olives, pitted and chopped
2 cloves garlic, minced
½ bell pepper, seeded and chopped
2 tablespoons capers
Salt
Freshly ground black pepper

Cut the baguette lengthwise, scoop out the moist center of the bread, break it into chunks, and place in a bowl. Drizzle olive oil over the scooped-out halves of the bread. Combine the tomatoes, onion, olives, garlic, bell pepper, and capers with the bread chunks, mixing well. Add salt and pepper to taste, and enough olive oil to thoroughly saturate the bread.

Fill one half of the bread shell with the mixture. Cover with the other half, squeezing the halves together; then wrap and refrigerate for half a day. Just before serving, remove from the refrigerator and cut into ½-inch-thick slices to serve as hors d'oeuvres or at luncheon.

❧ *BRUSCHETTA OR FETTUNTA* ❧

*I*talian friends have reprimanded me for calling this bread-soaked-in-olive-oil *bruschetta*. If I use Tuscan olive oil, they tell me, it should be called *fettunta*. A friendlier chap tells me that the exact same dish is called a *tostada* in his part of Spain. I've no doubt that wherever there is a new pressing of olive oil, olive farmers and mill operators sample their oil in just this way, whatever their appellation for the treat. Remember it is a wintertime dish, to be eaten by the fireplace.

> *A loaf of country-style, crusty white bread (not sourdough)*
> *Cloves of peeled garlic*
> *A cruet of the latest pressing of fine olive oil*
> *Salt*
> *Freshly ground black pepper*

Cut the bread into ½- to ¾-inch-thick slices. The traditional way to toast the bread is over a grill right in the fireplace. A toaster-broiler works just as well, but won't give the characteristic wood-fire taste. When the bread is golden crisp on both sides, serve a piece to each guest along with one or more cloves of garlic, cut in half. Everyone should then rub his toast with the garlic until it virtually disappears into the bread, and drizzle olive oil over the toast until it is completely saturated. Salt and pepper to taste. Serve with plenty of napkins.

❧ *MUFFALETTA SANDWICH* ❧

*S*ome epicures and bon vivants would rather vacation in New Orleans than anywhere else in the world, bar nowhere. New Orleans has its own cuisine and a long list of specialties, such as beignets, oysters Rockefeller, and, of course, muffaletta (the *e* is pronounced ŏ as in *pot*) sandwiches. Residents of New Orleans can't agree on which market makes the best; all the sandwiches are architectural accomplishments. There are similar sandwiches to be found around the

Mediterranean (see page 93, for example), but none is so voluptuous and decadent, so exquisitely large, so pungent, so oily, so devilishly impossible to bite into, as an extravagantly made muffaletta. To make the day before:

Olive Salad

3 large cloves garlic, crushed
1 cup pimiento-stuffed green olives, chopped
1 cup "black-ripe" olives or Kalamatas, pitted and chopped
½ cup roasted sweet peppers, cut into chunks
1 cup fine olive oil
3 tablespoons chopped fresh parsley
2 tablespoons white wine vinegar

Mix the garlic, olives, peppers (bottled or homemade), oil, parsley, and vinegar, and let stand overnight.

1 large, round, freshly baked Italian bread
⅓ pound salami or more, sliced
½ pound provolone, sliced
½ pound mild cheese (such as havarti), sliced
⅓ pound mortadella or more (or prosciutto or coppa)

To assemble sandwich: Cut the bread in half horizontally. Scoop out some of the center of the loaf and reserve for later use as bread crumbs. Drizzle olive oil from the olive salad on both halves of the bread. The bread should be saturated. On one half lay, in this order: salami, olive salad, provolone, mild cheese, and mortadella. Top with the other half of the loaf. Slice in wedges and serve with plenty of napkins.

Serves 6 hearty eaters.

✿ PANE CON OLIVE ✿
Bread with Olives

This crusty bread from Tuscany comes to us from a great Florentine cook, Giuliano Bugialli, who has added olives to make it "con olive." Allow about 6 hours to prepare and cool.

For the "sponge" (first rising)

*1½ ounces (3 cakes) of compressed fresh yeast
or 3 packages dry active yeast
¼ cup lukewarm water or hot water from the tap,
depending on the yeast
½ cup plus 1 tablespoon unbleached all-purpose flour*

For the dough (second rising)

*6 cups unbleached all-purpose flour
Pinch of salt
1¾ cups lukewarm water
1 teaspoon finely chopped fresh rosemary*
2 teaspoons olive oil
½ pound Kalamata olives, pitted*

Dissolve the yeast in the water in a small bowl, stirring with a wooden spoon. Place the ½ cup flour in a larger bowl, add the dissolved yeast, and mix with a wooden spoon until all the flour is incorporated and a small ball of dough is formed. Sprinkle the additional tablespoon of flour over the ball of dough, then cover the bowl with a cotton tea towel and put it in a warm place away from drafts. Let stand until the dough has doubled in size, about 1 hour.

Arrange the 6 cups flour in a mound on a pasta board, then make a well [it will be large-circumferenced and shallow]. Place the sponge from the first rising in the well, along with the salt, ½ cup of the lukewarm water, 2 teaspoons olive oil, and rosemary.

---

*I have added rosemary to Bugialli's recipe.

With a wooden spoon, carefully mix together all the ingredients in the well, then add the remaining water and start mixing with your hands, absorbing the flour from the inside rim of the well little by little.

Keep mixing until all but 4 or 5 tablespoons of the flour are incorporated (about 15 minutes), then knead the dough with the palms of your hands, in a folding motion, until it is homogeneous and smooth (about 20 minutes), incorporating the remaining flour, if necessary, to keep the dough from being sticky.

Add the olives to the dough and knead until they are evenly distributed. Lightly oil a 10-inch springform pan. Place the dough in the pan, cover with a cotton tea towel, and put in a warm place, away from drafts. Let the dough stand until doubled in size (about 1 hour); the time will vary a bit, depending on the weather.

Preheat the oven to 400°F.

When the dough has doubled in size, remove the towel and immediately place the pan in the oven. Bake the bread for about 60 minutes; do not open the oven for 30 minutes after you have placed the pan in the oven.

Remove the pan from the oven, allow to cool for 5 minutes, then open it, and transfer the bread to a pasta board.

To cool the bread, stand it on one of its sides, not lying flat. The bread must cool for at least 3 hours before it is at its best for eating, and the room where the bread cools must be very airy.

Makes 1 loaf.

❊ *PISSALADIÈRE* ❊

*T*his southern French version of pizza is quite a simple dish and is lovely as a first course. It is nothing more than an onion pizza, but the most common variations are topped with olives and are made, of course, with olive oil. The simpler the pissaladière is, the better. The black olive/anchovy version below can be modified by substituting bitter green olives and shrimp, or mushrooms and Niçoises. A hearty black olive for the recipe below is best, such as the Nyons.

1 package active dry yeast
¾ cup hot water from the tap
Olive oil
2 cups unbleached all-purpose flour
Salt
3 tomatoes, peeled, seeded, and cut into chunks
2 medium yellow onions, thinly sliced
8 anchovy fillets, rinsed
⅔ cup firm flavorful black olives, pitted

Dissolve the dry yeast in hot water. Add 1 tablespoon of olive oil. Add the flour little by little, stirring with a wooden spoon, and incorporating a sprinkling of salt, until the dough is a workable consistency. Turn the dough onto a well-floured board and knead until smooth. (You may have to add flour as you knead if the dough is sticky.) In an oiled bowl covered with a cotton tea towel, allow the dough to rise in a warm place until it has doubled in size, about an hour.

In 2 tablespoons of olive oil, sauté the tomatoes and onions until the onions are soft and the liquid from the tomatoes has evaporated.

Press with your fingertips or use a rolling pin to spread the dough onto a 12-inch-diameter pizza stone if you have one; or wrap a board completely with aluminum foil, oil with olive oil, and spread the dough on it. Make a lip around the edge of the pizza. Scatter the onions and tomatoes on it evenly. Make an attractive grid with the anchovies. Put the olives in the squares. Drizzle 2 or more tablespoons of olive oil over the pizza and bake at 450°F about 30 minutes, or until the crust is crisp and golden brown. Serve hot.

❧ *FOCACCIA* ❧

*T*here are two flat focaccia breads available in North Beach, the Italian neighborhood of San Francisco: a moist, springy focaccia, which is often topped with tomato sauce, and a lighter focaccia, sprinkled with herbs and made almost flaky by olive oil. A recipe for the latter follows. It must be served warm.

> *2 packages (2 tablespoons) active dry yeast*
> *1 tablespoon sugar*
> *1 cup hot water from the tap*
> *⅓ cup olive oil or more*
> *2 teaspoons salt*
> *⅔ cup warm water*
> *5½ cups all-purpose flour*
> *1 yellow onion*
> *Coarse (kosher) salt or ground sea salt*
> *Fresh rosemary, chopped*

In a large bowl, combine the yeast, sugar, and the 1 cup hot water and allow the mixture to sit until it bubbles and expands. Add the olive oil, the 2 teaspoons salt, and warm water, and stir. Little by little, add the flour, stirring to incorporate it after each addition. The dough should be stiff. Turn it onto a floured board and knead for 10 minutes. Place the dough in an oiled bowl, turning to coat all sides, and cover with a tea towel. Put the bowl in a warm, draft-free place, and allow the dough to rise to double its original size, about an hour.

Slice the onion; place in a small bowl, cover it with water, and set aside. When the dough has doubled in size, punch it down and turn it onto a floured board. Roll the dough out with a rolling pin to a size large enough to cover a baking sheet, about 12 by 18 inches. Transfer the bread to a greased baking sheet and let it rise for another hour.

Preheat the oven to 400°F. Drain the onion slices and dry. "Dimple" the dough all over with your finger and dribble some olive oil over the surface of the bread. Sprinkle with coarse salt, rosemary, and onion rings. Bake on the middle rack of the oven for 20 minutes. Cut into rectangles.

Serves 4 as a first course.

PIZZA

A true Neapolitan pizza is quite different from the pizza most Americans are used to. It is baked on tile, not in a pan, in a hardwood-fired brick oven, and is topped with fresh tomatoes, not a thick, sweet sauce. The effect is a crisper, more breadlike dough and a more subtle product overall.

To approximate an Italian oven, buy enough quarry tiles (available at building supply stores) to cover the middle shelf of your oven. The tiles must be heated at least 20 minutes at baking temperature before they are warm enough to receive the pizza. Use a wooden pizza paddle or the underside of a baking sheet covered with cornmeal to place the pizza in the oven. (The cornmeal will allow the pizza to slide off the paddle or baking sheet onto the tiles.) Or place the pizza in a pan on top of the tiles if you do not have a paddle or do not want to risk leaks. You will miss the benefits of the absorbent tiles, which make for a crisper crust, but the extra heat will make the center cook at the same rate as the edges of the pizza. If you have no tiles, use an old, non-laminated board, completely wrapped with aluminum foil and well-oiled with olive oil.

The following two recipes are an effort to please both tastes: an American pizza (complete with American olives) and a Neapolitan pizza. Use the following dough recipe for both pizzas.

Pizza dough

1 package active dry yeast
1 scant cup hot water from the tap
2½ cups all-purpose flour
1 tablespoon olive oil
Salt

In a bowl, dissolve the yeast in water. Use a fork to mix 2 cups of the flour with the dissolved yeast, olive oil, and a pinch of salt until it is pliable. Empty the dough onto a well-floured board and knead about 5 minutes, incorporating the remaining ½ cup flour. Place in an oiled bowl; cover with a clean, dry tea towel; put in a warm, draft-free place (such as an oven with a pilot light); and let the dough rise until it has doubled in size, at least 1½ hours.

Turn the dough onto a board or tray, cover with a damp tea towel, and let it rest at least half an hour. (At this point, you can refrigerate the dough overnight; be sure to allow at least 1 hour for the dough to warm to room temperature before forming a pizza.)

To form the pizza by hand, dust the board lightly with all-purpose flour. With the fingertips, press the dough into a circle of even thickness, about 7 inches in diameter. Pick up the circle between the palms and fingers of both hands and gently stretch the edge of the circle. Work around the edge, letting the dough hang. As the dough stretches to about 9 inches in diameter, drape it over the back of one hand and wrist and stretch it with the other hand. Continue working around the circle until the dough is about 16 inches across, or, if you are going to use a rectangular baking sheet, 12 by 18 inches.

If that method is too difficult to follow, the dough may be rolled out with a rolling pin; or, for a cruder looking pizza but one with a lighter dough, simply press the dough with your fingertips to the desired diameter.

Place the dough on an old, non-laminated board wrapped completely with aluminium foil, or on a baking sheet sprinkled with cornmeal; or, if cooking directly on tiles, place the dough on a pizza paddle sprinkled liberally with cornmeal (if you do not have a pizza paddle, use the underside of a baking sheet sprinkled with cornmeal). Pinch the edge of the dough slightly to form a raised rim.

❧ *PIZZA DÉCLASSÉ* ❧

This is the pizza Americans have become used to: It is made with a thick sauce and is covered with lots of cheese and, in this case, olives. Prepare the olives at least a day ahead of time.

Marinated Olives

6 ounces pitted "black-ripe" olives, drained and chopped
⅓ cup olive oil
1 tablespoon chopped fresh rosemary
1 clove garlic, minced
Zest of 1 lemon, grated
Freshly ground black pepper

Mix the above ingredients in a jar with a lid. Shake until all the ingredients are distributed among the olives. Turn the jar from time to time to redistribute the oil. Leave at room temperature overnight.

Tomato Sauce

2 28-ounce cans whole tomatoes, drained and chopped
½ tablespoon dried oregano
3 large cloves garlic, minced
Pinch of cayenne

Place the tomatoes in a large, heavy pot. Add the oregano, garlic, and cayenne and simmer rapidly, stirring often, until the mixture is the consistency of very thick sauce, about 1 hour.

Olives and their marinade
1 recipe pizza dough, prepared as on page 100
Sauce
½ pound fresh, whole-milk mozzarella, grated (about 2 cups)
8 rinsed anchovy fillets
⅙ pound Asiago, Parmesan, or Romano cheese, grated (about 1 cup)

To assemble: Remove the olives from their marinade. Brush the dough with the marinade, completely covering the surface of the dough, including the edges.* Spread the sauce on the dough; sprinkle with mozzarella; distribute the olives evenly; place the anchovies at even intervals; and top with the grated aged cheese.

Place the pan or board in the middle of a preheated oven, or slide the pizza onto the tiles in a preheated oven, and bake at 450°F until the dough is medium brown on the edges, 20 to 25 minutes.

Makes one 16-inch pizza.

❧ *NEAPOLITAN PIZZA* ❧

1 recipe pizza dough, prepared as on page 100
3 tablespoons olive oil
6 tomatoes, peeled, seeded, chopped, and well drained
4 cloves garlic, thinly sliced
¼ pound fresh, whole-milk mozzarella cheese, grated (about 1 cup)
1 tablespoon capers
6 anchovy fillets, rinsed (optional)
Salt
Freshly ground black pepper

Brush the pizza dough with the olive oil. Distribute the tomatoes, garlic, and cheese evenly over the pizza. Top with capers and, if desired, anchovies. Add salt and pepper to taste.

Bake in a preheated oven at 450°F for 20 to 25 minutes, or until the edges of the pizza are medium brown.

Makes one 16-inch pizza.

---

*If you want a crust with no trace of sogginess and you are cooking in a pan or on a board, melt half of the mozzarella over the crust in the oven before spreading the sauce, and reduce cooking time about 5 minutes.

First Courses
and Sauces

❧ *PASTA ALL'UOVO* ❧
Egg Noodles

Many Italians maintain that fresh egg pasta should be served with cream and butter sauces, whereas oil and tomato sauces are best with packaged, dried pasta. The following is a good, basic pasta recipe that can be served fresh after it is cut, as described here, or dried. I use a hand-cranked pasta machine to make the noodles.

1½ cups unsifted all-purpose flour
1 egg
1 egg white
1 tablespoon olive oil
1 teaspoon salt
Water

Put the flour in a mixing bowl and make a well in the center. Put the egg, egg white, olive oil, and salt in the well. With a fork, work all the ingredients into the flour. Form the mixture into a ball with your fingers. Add a little water to incorporate any remaining flour that will not stick to the dough, but keep the dough as dry as possible.

Knead the dough for about 10 minutes until it is shiny and elastic. Wrap it in waxed paper and let it rest for another 10 minutes. Cut the dough into thirds before running it through the pasta machine.

Start with the machine at the number 1 setting, dusting the dough with plenty of flour to keep it from sticking. Run the pasta through twice on each of the lower settings. After they are as thin as you want them, stack the pasta strips one on top of the other with plenty of flour between them.

Run the pasta through the cutter at the desired width. Put the freshly cut pasta in a large pot of rapidly boiling, salted water and stir with a fork once immediately after it is in the water. Boil for a *very* short time—as little as half a minute for the thinnest of pasta— until *al dente*.

Serves 4 as a first course or 3 as a main course.

PASTA WITH CREAM,
❧ FETA, OLIVES, AND BASIL ❧

Surprisingly, this rich first-course dish only starts the juices flowing. It should be followed by a light main course. Fresh basil is important to offset the salty feta and olives; dried basil does not substitute in this case. The basil and olives are added at the last minute so they will not discolor the sauce.

1 cup heavy cream
¼ pound feta cheese, crumbled
1 recipe egg noodles (page 105), very thinly rolled
Salt
⅓ cup dry-cured black olives, pitted and cut into small pieces
⅔ cup finely chopped fresh basil
Cracked peppercorns

In a wide skillet, reduce the cream by about a quarter; add the feta. Stir until the feta is somewhat melted, but the sauce is still lumpy.

Run the pasta through the pasta machine at the desired noodle width. Plunge the pasta into boiling, salted water and cook until *al dente*, considerably less than 1 minute. Drain in a colander, add to the sauce with the olives and basil, and toss.

Serve as a first course with cracked peppercorns to taste.

Serves 4.

❧ PASTA WITH PRAWNS AND PICHOLINES ❧

This is a delicate dish in spite of the heavy cream: The colors are pastels, the flavors refined. It is very nice with a good white Burgundy or a California Chardonnay.

4 tomatoes, peeled, seeded, and diced
Salt
2 recipes egg noodles (page 105)
1 tablespoon unsalted butter
2 cloves garlic, minced
Zest of 1 lemon, grated
1½ pounds prawns, peeled
1½ cups heavy cream
2 bunches fresh basil, chopped
1 cup picholine olives, pitted and cut into pieces
¼ pound Parmesan cheese, grated

Lightly salt the tomatoes and set aside to drain in a colander. Prepare the pasta, roll out, and cut, using plenty of flour so the strands do not stick together. In a large (5-quart) pot, melt the butter and cook the garlic until lightly browned. Add the lemon zest, prawns, and drained tomatoes. Cover and cook a few minutes until the prawns have just changed color. Remove prawns and set aside. Drain excess liquid. Add the cream to the pan and reduce the sauce by a third.

Cook the pasta in rapidly boiling, salted water until *al dente* (1 minute or less for fresh pasta), drain in a colander, and add to the sauce along with the prawns, basil, olives, and cheese. Toss and salt to taste.

Serve with plenty of freshly grated Parmesan cheese at the table.

Serves 8.

❋ *JULIE'S PUTTANESCA SAUCE* ❋

*T*wo stories: Prostitutes cook up this fragrant sauce to lure customers to their door; wives who spend their afternoons engrossed in *amours* can prepare this delicious dish quickly, keeping their husbands satisfied and unsuspecting.

4 large, peeled tomatoes
Salt
2 small hot red peppers, minced
4 tablespoons olive oil
4 large cloves garlic, crushed
6 anchovy fillets, rinsed and chopped
2 cups sliced Kalamata or ripe California olives
4 tablespoons capers, chopped
1 pound dried pasta
Fresh parsley, chopped

Slice the tomatoes (they should come to about 3 cups), salt them, and let them drain in a colander. Sauté the peppers in the olive oil 1 minute; then add the garlic and sauté for another minute. Add the anchovies, drained tomatoes, olives, and capers. Salt to taste. Simmer 10 minutes. Spoon onto the pasta of your choice and sprinkle with fresh parsley.

Serves 4.

❋ *POLENTA* ❋

Polenta—boiled cornmeal—is a perfect medium for olive oil. Served fresh just after it is made, it is a mild porridge over which to pour fruity olive oil with bits of savory things, such as clams, squid, mussels, or sausages, and garlic, or what have you. The next day, it will make a hearty breakfast or luncheon, sliced and fried in olive oil.

2 quarts water
2 teaspoons salt
1 pound polenta (coarse cornmeal)*

In a large pot, bring the water and salt to a boil. Add the polenta very slowly, stirring constantly with a wooden spoon. Cook, stirring all the while, until the polenta comes away from the sides of the pot, at least 25 minutes. When the polenta is cooked, turn off the heat and allow it to rest for a few minutes. Then turn it onto a wet wooden board. The polenta can be cut with a string and transferred with a spatula into shallow bowls, then topped with your choice of toppings and plenty of fine olive oil.

Serves 10 to 12.

---

*Some Italian cooks like to mix one-half fine to one-half coarse cornmeal for their polenta.

❧ *TRI-COLOR VEGETABLE TERRINE* ❧

*T*he original *terrine* was an earthenware dish, which gave its name to the concoction cooked in it. A vegetable terrine is a simpler dish than a meat terrine: There are no bards of bacon, no weights, no gelées.

This pretty orange, green, and cream-colored terrine is held together as much by the grape leaves as by the eggs; the leaves provide a contrasting flavor and texture as well as structural support. The dish makes a refreshing first course on a hot summer evening, served with tomato vinaigrette (page 122).

3 medium-size artichokes
Juice of ½ lemon
2 tablespoons olive oil
1 teaspoon salt

Wash and trim the tops and stems of the artichokes, cutting off the tips of the leaves and any tattered outer leaves. Cut in half. Rub with lemon juice to prevent discoloration. Place them cut-side down in the olive oil in a casserole large enough to accommodate the artichokes side by side. Add enough water to cover completely (about 6 cups); add salt; cover and simmer until tender, about 25 minutes. Leave in the liquid to cool.

4 tablespoons olive oil
4 large carrots, peeled and sliced
2 cups water
1 teaspoon salt
1 egg
⅛ teaspoon ground nutmeg
1 tablespoon cream

With 3 tablespoons of the olive oil, boil the carrots in the salted water until just tender, about 12 minutes. Plunge them into cold water to refresh, and drain. In a blender or food processor, blend the drained carrots with the egg, nutmeg, cream, and the remaining 1 tablespoon olive oil. Place in a bowl and set aside.

3 cups tight broccoli flowerets, firmly packed
½ teaspoon salt
1 egg
1 clove garlic, crushed
1 tablespoon olive oil
1 tablespoon cream
1 teaspoon chopped fresh thyme or ½ teaspoon dried oregano

Plunge broccoli into a large pot of boiling water and boil until just tender, about 4 minutes. Refresh in cold water and drain. In a food processor or blender, blend drained broccoli with salt, egg, garlic, olive oil, cream, and thyme or oregano. Place in a bowl and set aside.

1 small head cauliflower
½ teaspoon salt
1 egg
1 tablespoon olive oil
1 tablespoon cream

Trim and rinse the cauliflower; cut off stems, break into pieces, and steam until just tender, about 10 minutes. Refresh in cold water and drain. In a food processor or blender, blend the cauliflower, salt, egg, olive oil, and cream. Hold in processor (or blender) container.

7 prepared grape leaves or parboiled fresh grape leaves
4 tablespoons olive oil
½ cup firm Kalamata olives, pitted and chopped
Watercress or lettuce
Tomato vinaigrette (page 122)

(Continued on page 114.)

Rinse the grape leaves and pat dry. Grease a 9- by 4- by 3-inch terrine, Pyrex bread pan, or similarly shaped enamel pan with olive oil and line the bottom and part way up the sides with overlapping grape leaves. Pour the carrot mixture into the pan. Remove the chokes and leaves from the artichoke hearts (saving the leaves to dip in vinaigrette). Arrange the hearts in a row over the carrot mixture. Arrange the rest of the grape leaves so they adhere to the inside of the terrine, and drape over the sides. Pour the broccoli mixture over the carrots and artichokes. Sprinkle the olives evenly over the broccoli mixture. Pour the cauliflower mixture over the olives. Fold the grape leaves over the top, tearing off the excess where two or more layers of grape leaves overlap. Drip olive oil over all to keep the grape leaves from drying out. Bake in a boiling water bath at 350°F for 1 hour. Let cool.

Refrigerate until the terrine is thoroughly cold. Carefully turn onto a serving platter and cut with a sharp serrated knife, being careful to cut cleanly through all the grape leaves. Serve the slices on individual beds of watercress or lettuce, and spoon the tomato vinaigrette over all.

Makes 8 slices.

❊ *PAPPA AL POMODORO* ❊

"You've got to have a recipe for pappa al pomodoro," my Tuscan friend insisted. And so I do. The dish should be served with plenty of good olive oil at the table to pour over it. Although it has the consistency of pudding, it can be substituted as a soup course; it is wonderful served as a simple lunch.

6 cloves garlic, chopped
Hot red pepper flakes
⅓ cup olive oil
2 pounds ripe tomatoes, peeled, seeded, and chopped
1½ pounds stale, dried bread
(a crusty whole wheat or a tasty, crusty white)
4½ cups chicken stock
1 bunch fresh basil leaves
Salt
Freshly ground black pepper
A cruet of fine olive oil

In a 5-quart covered pot, sauté the garlic and a hearty shake of hot pepper flakes in the olive oil. Add the tomatoes and simmer 15 minutes. Cut the bread into 1½-inch cubes. Then add the bread, chicken stock, and basil leaves, and salt and pepper to taste. Stir very well, so that all the liquid is absorbed by the bread. Simmer over low heat for 10 minutes, covered. Then turn off the heat and let the pappa rest for an hour.

Reheat and serve with plenty of fine olive oil.

Serves 8.

❧ *RISOTTO CON SALSICCE E FUNGHI* ❧
Rice with Sausage and Mushrooms

Risotto is a rich first course typical of northern Italian meals. It is a wonderful way to taste the finest olive oils and it can be as addicting as pasta. The sauce may be varied by adding mussels or clams and steaming them before the oil is added to the sauce.

Do not expect a light, fluffy rice; risotto is heavy, and the rice kernels adhere to one another.

Risotto

3 cups strong chicken stock
1 large yellow onion, finely chopped
4 tablespoons olive oil
4 tablespoons unsalted butter
1⅓ cups raw Italian Arborio rice (superfino)
⅓ cup grated Parmesan cheese, tightly packed
Freshly ground black pepper
Fine Tuscan olive oil

Put the chicken stock on the stove to simmer. In a heavy casserole, sauté the onion in oil and butter over a medium heat until golden. Add the rice and mix well with a wooden spoon until the rice is well coated with butter and oil. With a ladle, add some of the stock, stirring the rice constantly until it has absorbed the stock. Add more stock as you continue to stir. Continue the procedure, always waiting until the rice has absorbed all the stock before adding more. When all the stock has been added in this manner, the rice should be *al dente*; it will look much soupier than what we Americans are used to. If the rice is too chewy, add more broth or a little water and stir it in. A proper risotto takes a little patience, and constant stirring.

Remove from the heat and stir in the cheese and pepper. Serve in individual bowls as a first course plain (with only fine Tuscan olive oil poured over it), or with the sauce that follows.

Sausage and Mushroom Sauce

½ cup dried mushrooms*
4 small sweet Italian sausages
3 cloves garlic, minced
⅔ cup fine olive oil
3 large tomatoes, peeled and seeded
Hot red pepper flakes

Soak the mushrooms in warm water for at least an hour. Squeeze out the excess water and slice. In a skillet, sauté the sausages and garlic in 2 tablespoons of the oil, adding the tomatoes when the sausages are almost done. Stir frequently. When the tomatoes are cooked, cut the sausages into small pieces, add the mushrooms, pour in the rest of the oil, and season with the hot pepper flakes to taste. Place over a low heat to warm the sauce, stirring constantly, being careful not to cook the oil as cooking changes its flavor. Pour over the risotto.

Risotto must be served immediately or it will become gummy. It does not reheat well.

Serves 4.

---

*South American dried mushrooms are flavorful and inexpensive.

GREEK FISH SALAD
❧ WITH AVGOLEMONO SAUCE ❧

This cool, delicate salad makes a refreshing hot-day lunch or an elegant first course. Use halibut, swordfish, or other firm-fleshed fish.

3 tablespoons olive oil
2 cups water
2 bay leaves, broken
6 whole peppercorns
1 small yellow onion, peeled and sliced
1 small lemon, sliced
2 pounds fish, in thick fillets or steaks

In a frying pan with a lid, simmer the oil, water, bay leaves, peppercorns, onion, and lemon for 5 minutes. Then add the pieces of fish side by side, cover, and cook for 5 to 8 minutes, depending on the thickness of the fish. (Check with a fork after 5 minutes and remove from the heat when just cooked.) Remove the fish carefully from the liquid and let cool; then refrigerate. Over high heat, reduce the liquid in the pan to ½ cup.

Avgolemono Sauce

2 egg yolks
½ cup hot fish stock (from above)
1½ tablespoons fresh lemon juice
½ cup sour cream
½ teaspoon salt

In the top of a double boiler over simmering water, whisk the egg yolks with a wire whisk and slowly add the hot fish stock. Add the lemon juice and cook, whisking constantly, until the sauce thickens. Remove from the heat and mix in the sour cream and salt to taste. Correct seasoning and refrigerate.

Cold fish, cut into serving pieces
1 cucumber, sliced
12 cherry tomatoes, halved
Avgolemono sauce
24 Greek olives, Kalamatas or large cracked greens

To assemble: Arrange the fish on each plate with the cucumber slices and tomatoes. Pour the Avgolemono sauce over all. Garnish with the olives.

Serves 8.

✺ *CARPACCIO* ✺

This elegant and simple first course is a fine way to enjoy the best olive oil: The unassertive raw beef will not compete with the oil, and the capers will heighten its flavor. With luck you can find good wild mushrooms at your produce store, or a few imported *porcini* or *cèpes*.

2 ounces fillet of beef per person, sliced paper thin
Fine olive oil
Fresh mushroom(s), sliced with a vegetable peeler
1 teaspoon capers per person
1 lemon wedge per person

Pound each slice of beef between two pieces of waxed paper with the flat side of a meat pounder, or with a broad, heavy knife. To serve, drizzle a thin coating of olive oil on each serving plate. Arrange the meat on each plate artfully and garnish with the thin mushroom slices, capers, and lemon wedges. Pass additional oil in a cruet, letting each guest add oil to taste. Hide the salt shakers for this course.

❧ AÏOLI OR AILLOLI ❧

*O*ne of the loveliest, most festive meals I ever had was an *aïoli monstre*, modeled after the town-wide garlic feasts held in Provençal villages. At a groaning table were parboiled, steamed, or raw vegetables of all varieties (fennel bulb, broccoli, spring potatoes, cherry tomatoes, mushrooms, asparagus, radishes, artichokes, carrots, cauliflower, and so on), hard-boiled eggs, chunks of poached salmon, and thinly sliced rare beef. We dipped it all in a twelve-clove aïoli, and drank a young Bordeaux. I smelled like Provence for three days afterward.

A light olive oil is most pleasing to the palate for this potent sauce; a gold- rather than green-hued oil is most pleasing to the eye. Start this recipe with 4 cloves of garlic, then taste. Add more mashed garlic, according to your preference and the strength of your garlic. A true aïoli will have about 8 cloves of garlic to 1 cup of olive oil.

4 to 8 cloves fresh sweet garlic
1 tablespoon dry bread crumbs
White wine vinegar
1 egg yolk
1 cup fine olive oil
Salt

In a bowl with a curved bottom, mash the peeled garlic cloves with a pestle. Moisten the bread crumbs with vinegar and work into the garlic paste. With a wire whisk, whip the egg yolk with the garlic paste for a few minutes, until it is a pale yellow. Then add the oil, drop by drop, continuing to whisk. After ⅓ of the oil is incorporated, increase the rate of oil addition to a very fine drizzle (still whisking all the while), until all the oil is incorporated. Salt to taste as the last of the oil is beaten in. The aïoli should have the consistency of a thick mayonnaise.

❧ *MAYONNAISE* ❧

*T*he better the olive oil, the better the mayonnaise. If you are going to flavor the mayonnaise heavily, however, or use it with piquant ingredients, you can use a lesser quality olive oil or mix the olive oil with an equal part peanut oil. In fact, some cooks prefer to cut the flavor of the olive oil with a neutral oil.

2 egg yolks
¼ teaspoon dry mustard
½ teaspoon salt
2 tablespoons white wine vinegar
1 cup fine olive oil
2 tablespoons lemon juice

Have all ingredients at room temperature. Beat the egg yolks with a wire whisk until they are pale yellow. Beat in the mustard, salt, and ½ teaspoon of the vinegar. Beat in, a drop at a time, ½ cup of the olive oil. In a separate container, combine the remaining vinegar and the lemon juice. Alternate adding this mixture, to taste, to the mayonnaise with the remaining ½ cup olive oil, a few drops at a time, whisking all the while.

The bitterness of the mustard can be alleviated by adding a pinch of confectioner's sugar at the same time the mustard is added, or by using a pinch of cayenne pepper instead of mustard (or omit the spice altogether).

Mayonnaise is highly perishable and must be kept refrigerated if not used immediately.

Makes 1⅓ cups.

✿ BASIC TOMATO VINAIGRETTE ✿

Although the best salads are made simply, with fine olive oil, fine vinegar, the freshest greens, salt, and perhaps a little pepper, I include the following recipe because it goes so well with slices of the vegetable terrine (page 110). Combine in a food processor or blender:

3 fresh tomatoes, peeled and seeded (or 3 canned tomatoes)
4 tablespoons olive oil
2 tablespoons white wine vinegar
Pinch of salt
1 clove garlic, finely minced
Dash of Tabasco sauce

❋ *PESTO* ❋

*P*esto has become a standby in many American households. Its marvelous flavor comes mainly from basil—one of the world's great garden herbs—and fine olive oil. Too much garlic added to this sauce tends to overpower the basil. Some connoisseurs of pesto contend that one should make it entirely without garlic to better savor each subtlety.

2 cups fresh basil leaves
1 teaspoon salt
Freshly ground black pepper
4 cloves garlic, minced
2 tablespoons pine nuts
1½ cups fine olive oil
½ cup freshly grated Parmesan or Romano cheese

Chop the basil very fine; then put it into a mortar and crush with a pestle until it is almost a paste. In a bowl, work the salt, pepper to taste, garlic, and pine nuts into the basil paste with the pestle, crushing the pine nuts as you pound. Add the olive oil a little at a time, continuing to crush the mixture with the pestle as you add the oil. When the sauce is a liquid rather than a paste, add the grated cheese.

Serve with hot pasta or as a flavoring in sauces or soups. You can also freeze a portion of it for later use.

Makes 1½ cups.

Salads

❋ GREEK, OR RURAL, SALAD ❋

This salad can be infinitely varied according to what is in season and your tastes and preferences. Many think it must include spinach, and, indeed, spinach goes very well with the feta cheese. You can omit most of the greens in favor of fresh herbs, or use lettuces with abandon, especially during the winter. Consequently, you determine the amount of each ingredient. A purple onion, thinly sliced, is wonderful with the anchovies and feta. A good rule of thumb for feta cheese is to allow ⅛ pound per person. A ratio of 4 parts olive oil to 1 part vinegar is appropriate for this salad.

Lettuces, such as romaine, red-leaf, curly endive, and so on
Spinach
Purple onion, thinly sliced
Rocket (also known as rocquette or rugola)
Watercress
Fresh herbs, such as mint, cilantro, basil, thyme, and so on
Cucumber, thinly sliced
Ripe, brine-cured Greek olives
Feta cheese, crumbled
Capers
Dried oregano (rubbed between the palms and used copiously)
Garlic cloves, pressed
Salt
Freshly ground black pepper
Anchovy fillets, rinsed
Fine olive oil
Red wine vinegar

Combine the ingredients for your salad in a large bowl and toss. Serve the salad with a hearty, country-style bread and unsalted butter.

✵ SALADE NIÇOISE ✵

There are no strict recipes for this famous salad, for the ingredients depend on what vegetables are in season; choose a few from each category below. Certain ingredients are a must: first-rate olive oil, red wine vinegar, garlic, and olives, preferably Niçoises or Niçoise-type (small, black, and brine-cured). I think of this wonderful salad as having tuna, even if it is canned.

Lettuces and other greens

bibb
red leaf
curly endive
butter
romaine
spinach
watercress
rocket (also known as rocquette or rugola)

Raw vegetables

summer squashes
radish
radicchio
scallions
tender green peas
fennel bulb
tomatoes

Parboiled vegetables

asparagus
broccoli
carrots
cauliflower
green beans
bell pepper

Boiled or steamed items

potatoes
hard-boiled egg
tiny artichokes (trimmed, rubbed with lemon juice, and steamed)

The arrangement of your *salade Niçoise* can make it special, and worthy of a whole meal. Start with a selection of lettuces. Dress them together; toss; then arrange them on a large serving platter. Next, slice the raw vegetables as appropriate (thin, whenever possible), fan them, and arrange atop the lettuce to one side, leaving the center for the tuna. Arrange the parboiled vegetables, sliced, if appropriate, and fanned. Arrange the potatoes in a small mound, and fan the sliced hard-boiled egg. Then add the tuna in the center in large chunks if fresh (poached and cooled), or in the shape of the can if canned. Add salt and freshly ground pepper to taste. Dress with a vinaigrette made with crushed garlic, red wine vinegar, and the best olive oil you have. Garnish with olives, capers, anchovies, etc.

❧ ONION, OLIVE, AND LETTUCE SALAD ❧

This simple salad is very refreshing on a hot summer afternoon or evening. The onions, "cooked" a little in the dressing, are tame enough not to overpower the rest of the meal.

¼ cup fine olive oil
3 tablespoons good cider vinegar
¼ teaspoon salt
Freshly ground black pepper
1 large purple onion, very thinly sliced
1 cup Kalamata olives
1 bunch watercress
1 bunch romaine lettuce
½ bunch bitter greens (escarole or chicory, for example)

Pour olive oil, vinegar, salt, and pepper to taste over the sliced onion and let stand half a day, tossing from time to time. Add olives and toss again. Wash the watercress and discard the tough stems. Wash the romaine and bitter greens; dry them, and tear into pieces. Mix the olives and onions with the greens and toss.

Serves 6.

❧ PARSLEY AND OLIVE SALAD ❧

Although this rich salad is a fine accompaniment for Italian cold cuts and good Tuscan or sourdough bread, it is versatile and can also serve as a condiment at the table with various hot dishes, such as grilled chops or stews. It does not keep well.

1 cup pimiento-stuffed green olives, chopped and firmly packed
1 cup freshly toasted walnuts, chopped
2 cups parsley, finely chopped
3½ tablespoons fine olive oil
1 tablespoon lemon juice
½ tablespoon capers, chopped
1 large clove garlic, crushed
Tabasco sauce

Combine olives, walnuts, parsley, olive oil, lemon juice, capers, and garlic. Season with Tabasco sauce to taste. Mix thoroughly. Serve at room temperature.

Serves 6.

❄ *PASTA AND OLIVE SALAD* ❄

Many varieties of pasta are available now, and their beautiful shapes make handsome salads. But I prefer the tiny, rice-shaped pasta for this salad because more surface area is covered by the dressing than with the larger types—an important factor with cold pasta.

This dish goes along well on picnics, since it contains no immediately perishable ingredients. Or it can be placed on a buffet table with cold cuts and other salads.

2 cups tiny, rice-shaped pasta (rosamarina or semini di melo)
3 tablespoons fine olive oil
1 tablespoon sherry wine vinegar
1 large clove garlic, crushed
⅓ cup Sicilian-style olives, pitted and finely chopped
¼ cup finely chopped roasted sweet red peppers*
2 scallions, finely chopped
¼ cup finely chopped parsley
¼ teaspoon sugar
Pinch of dried thyme

Plunge the pasta into a large pot of boiling, salted water. Stir and cook until just tender. Pour into a colander, rinse with cold water, and let drain a few minutes.

In a salad bowl, mix the pasta with the olive oil, vinegar, garlic, olives, red peppers, scallions, parsley, sugar, and thyme.

Serves 4 to 6.

---

*Bottled or canned roasted peppers can be found at most Italian delicatessens.

❧ TUNA, RICE, AND OLIVE SALAD ❧

This rich salad is perfect for a buffet with cold meats and a variety of other salads and pickles, breads, and cheeses. Serve at room temperature, but be sure to refrigerate it if you are preparing it ahead of time.

2 cups short-grain brown rice
Salt
1 recipe mayonnaise (page 121)
3 cloves garlic, crushed
1½ cups Sicilian-style green olives, pitted and coarsely chopped
2 celery hearts, diced
6½-ounce tin water-packed tuna, drained
Freshly ground black pepper
2 garden-fresh tomatoes
2 hard-boiled eggs
8 anchovy fillets, rinsed
½ bunch Italian parsley, chopped

Into 4 cups boiling, salted water, pour the rice and stir. Simmer, covered, until the rice is tender but chewy, about 35 minutes. Add more water if necessary. Pour the rice into a sieve, rinse with cold water, and let drain.

In a mixing bowl, thoroughly mix the drained rice, ¼ teaspoon salt, mayonnaise, garlic, olives, celery hearts, tuna, and pepper to taste. Turn into a serving bowl and garnish with sliced tomatoes, hard-boiled eggs, and rolled anchovy fillets. Sprinkle with parsley.

Serves 8.

❊ PEAR, WATERCRESS, AND OLIVE SALAD ❊

Juicy, sweet pears and a peppery olive oil are equally balanced in this salad, which is especially good following a heavy main course.

1 large ripe pear, or 2 smaller ones, peeled and cored
1 bunch watercress, washed, stems removed
¼ cup tiny Niçoise or Ligurian olives
2 tablespoons fine olive oil
Juice of ½ lemon
Pinch of salt

Cut the pear(s) into ½-inch cubes. Arrange the watercress on four serving plates. In a bowl, toss the pear pieces with the olives, olive oil, lemon juice, and salt. Mound the salad on the beds of watercress and pour any dressing remaining in the bottom of the bowl over the watercress.

Serves 4.

❋ *ARMENIAN EGGPLANT SALAD* ❋

*T*his salad is best freshly made and cooled to room temperature.

> *1 eggplant*
> *Salt*
> *Olive oil*
> *1 small yellow onion, finely chopped*
> *½ bunch parsley, finely chopped*
> *1 tablespoon white wine vinegar*
> *½ teaspoon cinnamon or more*

Cut the eggplant into medium-size chunks, skin and all. Salt the pieces, put into a bowl, and weight them with a heavy plate. Drain after they have sweated for an hour or more.

Sauté the drained eggplant in a heavy skillet with plenty of olive oil. Add oil as necessary to keep the eggplant from sticking. Leave the pieces unturned until they are browned a bit; then turn and cook until most sides are slightly browned. (The eggplant will be done when it no longer absorbs oil.)

Mix the eggplant in a bowl with the onion, parsley, vinegar, and cinnamon.

Serves 6.

Soups

❀ *GAZPACHO* ❀

*T*his unusual gazpacho with cream is a memorable soup I had years ago in Barcelona. Its flavors are so pronounced that it creates a dilemma—what to serve for a second course. It is difficult to find a better follow-up than a simple grill.

½ bell pepper
1 medium-size yellow onion, chopped
1 clove garlic, crushed
4 tomatoes, peeled, seeded, and coarsely chopped
1 celery heart, chopped
1 cup heavy cream
2 cups light chicken broth
2 teaspoons salt
¼ cup white wine vinegar
¾ cup fine olive oil
5 dashes Tabasco sauce or to taste
Cilantro

Blend the bell pepper, onion, garlic, tomatoes, and celery heart with the cream, chicken broth, salt, vinegar, ½ cup of the olive oil, and Tabasco sauce in a blender or food processor. (This may have to be done in batches, depending on the capacity of the machine.) Transfer to a large bowl and refrigerate until nicely cooled. Just before serving, mix in the last ¼ cup olive oil. Garnish each bowl with cilantro leaves.

Serves 6.

MINESTRONE FROM CHIANTI,
❧ COUNTRY STYLE ❧

This soup is best the day after it is made. If you want to skip the first day's version altogether, simply put two thick slices of bread into the pot of soup before it is refrigerated; they will thicken the soup nicely. The next day's version, *ribollita* (reboiled), should be heated until just simmering. The olive oil and freshly grated Parmesan cheese garnishes are essential both days.

1½ cups dried small white beans
5 cups water
1 large ham hock
⅓ bunch celery, chopped
1 potato, diced
1 onion, thinly sliced
1 bunch fresh parsley, minced
1 carrot, sliced
6 cloves garlic, minced
4 tablespoons olive oil
1 bunch Savoy cabbage, rinsed and coarsely chopped
1 bunch kale, rinsed and coarsely chopped*
3 large tomatoes, peeled
Salt
Freshly ground black pepper
1 loaf country-style bread, preferably Tuscan style
Freshly grated Parmesan cheese
Cruet of your best Tuscan olive oil

Rinse the beans and soak overnight in water to cover; then drain. In a covered pot, simmer the beans, water, and ham hock 1½ hours, or until the beans are tender. Remove from the heat.

In a large kettle, sauté the celery, potato, onion, parsley, carrot, and garlic in olive oil for 10 minutes, stirring often. Add the greens. Cut the tomatoes into coarse chunks and add them, with their seeds

---

*You may also use green-leaved Swiss chard, collard greens, or mustard greens.

and liquid, to the pot. Remove the skin from the ham hock and cut the meat from the bone. Add both the meat and the bone to the pot. With a slotted spoon, remove half the beans from their liquid and set aside. With a potato masher, mash the rest of the beans with their liquid and add to the pot.

Cover and cook 15 minutes. Add water to cover the ingredients (about 6 cups), cover, and simmer half an hour. Add the whole beans and season to taste, depending on the saltiness of the ham.

If you are serving the soup the day it is made, put a thick slice of bread at the bottom of a tureen. Add a few ladlefuls of soup, then another layer of thick bread, and another few ladlefuls of soup, until the tureen is full, topping with the soup. Sprinkle cheese on top, cover, and let sit 10 minutes. Serve in large bowls, and have oil and more cheese handy at the table. Do not scrimp when pouring the olive oil into the bowls.

Serves 6.

BEAN SOUP FROM TUSCANY

This delicious wintertime soup is easy to make, and goes nicely with any coarse, country-style bread.

1 pound dried small white or lima beans
1 quart water
1 quart light chicken stock
4 bay leaves
2 teaspoons dried oregano
6 tablespoons olive oil
2 teaspoons salt
¼ cup finely chopped fresh parsley
4 cloves garlic, crushed
5 tablespoons lemon juice
Cruet of your finest olive oil

Soak the beans overnight. Drain them and cover with water and chicken stock in a large iron or enameled pot. Add the bay leaves, oregano, and 6 tablespoons olive oil. Bring to a boil, cover, and simmer until the beans are tender, about an hour.

Add the salt. Purée half the beans in a blender or food processor with half the liquid, and mix with the whole beans and the remainder of the liquid. Add the parsley, garlic, and lemon juice, and simmer 10 minutes more. Serve in large soup bowls.

Encourage your guests to use the olive oil liberally.

Serves 4 to 6.

❧ *PAPAS VUIDA* ❧
Potato and Olive Oil Soup

*H*ere's another recipe by Anzonini del Puerto, the warm and wooing Spanish gypsy who gives to his friends through his piquant cooking and his Flamenco music. This soup is a perfect example of how Anzonini uses olive oil to substitute for meat and meat broth and supply a special richness. The soup is best served the next day, reheated.

6 cups water
2 tomatoes, peeled, seeded, and chopped
½ bell pepper, seeded and broken into pieces
1 large yellow onion, chopped
8 cloves garlic, peeled and left whole or coarsely chopped
1 cup olive oil
3 bay leaves
10 peppercorns
3 large unpeeled potatoes, sliced thick
Salt

In a large pot, bring the water, tomatoes, bell pepper, onion, garlic, olive oil, bay leaves, and peppercorns to a boil. Cover and simmer 45 minutes. Add the potatoes and simmer until tender, at least 45 minutes. The longer it is cooked, the better it becomes. Salt to taste.

Serves 8.

❧ OLIVE AND CHICK PEA SOUP ❧

Chick peas are seldom used in the United States, making only an occasional appearance—usually from the can—in salads and antipasto plates. I find the dried chick pea has a better flavor and texture than the canned. This recipe is Moroccan in inspiration, and is the closest I could come to finding an olive soup.

1 ½ cups dried chick peas
⅓ cup unsalted butter, room temperature
⅙ pound Kalamata olives, pitted and chopped
6 cups medium chicken stock
1 large unpeeled potato, diced
1 yellow onion, diced
Hot red pepper flakes
½ teaspoon ginger
½ teaspoon turmeric
Large pinch crumbled saffron threads
½ teaspoon cinnamon
2 teaspoons salt
Juice of 1 lemon
1 cup chopped parsley

Soak the chick peas overnight and boil until tender, at least 1 hour. Check the cooked chick peas and, if the skins seem to peel off easily, remove them; otherwise, don't bother.

Cream the butter with a wooden spoon and mix in the olives. Form a ball with the butter and olive mixture, cover in plastic wrap, and refrigerate.

In a blender or food processor, purée the chick peas with a little of the chicken stock until almost smooth. In a large pot, mix the puréed chick peas with the remaining chicken stock, potato, onion, hot pepper flakes to taste, ginger, turmeric, saffron, cinnamon, salt, and lemon juice. Simmer, uncovered, for an hour; then add the parsley, cover, and simmer for an hour more. Adjust the seasoning.

Serve hot, with a dollop of the cold butter and olive mixture melting in the center of the bowls.

Serves 6.

Vegetables

SQUASH GRATIN
🌟 WITH GARLIC AND OLIVE OIL 🌟

A gratin is any dish cooked in such a way that it forms a nice crust on top. This can be accomplished in a variety of ways—by putting the dish under a broiler, especially with a topping of bread crumbs or a dry grated cheese; or, as in this recipe, by baking uncovered for a considerable time.

About 3⅓ pounds butternut or other firm, orange squash
8 cloves garlic, finely minced
1 large bunch parsley, finely chopped
5 tablespoons flour
Salt
Freshly ground black pepper
⅔ cup olive oil

Peel and seed the squash, and cut it into ⅓-inch cubes. In a large bowl, toss the squash with the garlic and parsley. Then sprinkle in the flour, and salt and pepper to taste, tossing until all the cubes are coated with flour.

Grease a large gratin dish or a wide, shallow casserole with a little of the olive oil. Add the squash. Dribble the remaining olive oil over the squash in a crisscross fashion. Bake in a 350°F oven at least 2 hours, until the top of the squash has formed a crisp, brown crust.

Serves 8.

❧ ARTICHOKES À LA GRECQUE ❧

This seasonal dish can be made ahead of time if you remove it from the refrigerator long enough before serving for the oil to liquify. It needs the shiny black olives to liven its somewhat drab colors.

8 medium artichokes
½ cup fresh lemon juice
Salt
8 small carrots, peeled and cut in half lengthwise
2 unpeeled potatoes, cut into ½-inch cubes
1 cup olive oil
2 tablespoons chopped fresh dill
2 bunches scallions
24 plump, dry-cured Greek or California olives

Wash and trim the artichokes. Cut them in half and remove the chokes; then rub them with a little of the lemon juice and submerge in cold, salted water to prevent their discoloration. In a large frying pan (not iron) with a lid, or in a Dutch oven, place the artichokes flat-side down with the carrots, potatoes, olive oil, dill, salt to taste, the remaining lemon juice, and enough water to almost cover the artichokes. Cover and simmer until the artichokes are almost cooked, about 20 minutes. Add the scallions and cook 5 minutes more.

Transfer to a serving platter and allow to cool (you may refrigerate it if you prefer, but take it out before the oil congeals). Pour the sauce from the pan over the vegetables, and serve garnished with the olives.

Serves 8.

❧ SAVORY FENNEL ❧

*F*ennel, with its faint licorice flavor, has such a nice, crisp texture and delicate flavor I wonder why it isn't seen more often as an accompaniment for chops, roasts, and other meats, and as an ingredient in salads. Its only flaw is an aesthetic one: It is a pale, translucent green when cooked, and requires a bit of color to accompany it on a plate.

5 tablespoons olive oil
1 large yellow onion, thinly sliced
1 tablespoon chopped fresh rosemary
2 tablespoons toasted bread crumbs*
Salt
3 large fennel bulbs, washed and trimmed
1 cup dry white wine
Freshly ground black pepper
Fennel seeds, crushed

Heat 2 tablespoons of the olive oil in a large, flat skillet and cook the onion and rosemary until the onion slices are tender. Add the bread crumbs and cook the mixture until golden. Sprinkle with salt. Remove from the pan and set aside. Wipe the pan clean. Add the remaining 3 tablespoons olive oil.

Cut the fennel bulbs lengthwise and place flat-side down in the oil and cook over low heat, covered, until slightly browned, about 10 minutes. Add the wine and season with salt and pepper to taste. Cover and cook until the fennel is tender, about 20 minutes. Remove the fennel to a serving platter, reserving the liquid in the pan. Place the fennel in a warm oven.

Reduce the liquid to a thick consistency. Pour over the fennel. Return the onion mixture to the pan and reheat 1 minute. Top the fennel with the onions. Garnish with fennel seeds and serve.

---

*To make bread crumbs: In a blender or food processor, grind slices of good, dry bread (either stale or crisped—not browned—in the oven). Spread the bread crumbs in a thin layer and toast in a toaster oven until golden brown.

❧ *RATATOUILLE* ❧
Provençal Eggplant Stew

This popular and aromatic eggplant stew is time-consuming to make, but cooking the vegetables separately preserves the individual textures and flavors and is essential to its success. You will need a covered casserole, a skillet, and three bowls.

1 medium eggplant
Salt
3 medium zucchini
Olive oil
Freshly ground black pepper
2 yellow onions, thinly sliced
2 red bell peppers, seeded and cut into thin strips
4 cloves garlic, minced
4 ripe tomatoes, peeled, seeded, and chopped
½ cup chopped cilantro

Peel the eggplant, slice into ½-inch-thick rounds, then slice into 2- by 1-inch rectangles. Put the eggplant into a bowl and toss with salt. Cut the zucchini into ¼-inch slices, put into another bowl, and toss with salt. Let the vegetables sweat for half an hour, then dry the eggplant and zucchini on a towel.

In a skillet, sauté the eggplant in ¼ cup olive oil until lightly browned, adding more olive oil if necessary. Season to taste with pepper. Transfer to a bowl and set aside. In the same skillet, sauté the zucchini in another ¼ cup olive oil until lightly browned, but not limp. Season to taste with pepper; transfer to another bowl, and set aside.

In the same skillet, sauté the onions, red bell peppers, and garlic in 2 tablespoons olive oil until they are tender. Season to taste with salt and pepper; transfer to another bowl, and set aside.

In the same skillet, cook the tomatoes in 1 tablespoon olive oil until the liquid they exude evaporates, stirring occasionally.

In the casserole, layer half the eggplant, half the zucchini, half the red bell pepper and onion mixture, half the cilantro, and half the tomatoes. Repeat the layering with the remaining ingredients.

Cover and simmer 10 minutes. Uncover, and with a bulb baster, baste the stew with juices from the bottom of the casserole. Simmer, uncovered, until the juices are absorbed, about 10 minutes more, being careful not to singe the bottom. Serve hot, or cold as a salad or antipasto.

Serves 4 to 6.

Main Courses

❦ BACCALÀ ❦

*I*t is the dried, salted fish that gave Italian grocery stores the smell
I disliked as a child. So it came as a surpise to me that not only are
such fish expensive, but they are worth every penny of the cost. The
following stew is very good indeed. You may use any brine- or oil-
cured black olive.

> *2 pounds dried salt cod*
> *½ cup olive oil*
> *6 cloves garlic, minced*
> *5 large yellow onions, diced*
> *3 bell peppers, seeded and torn into pieces*
> *8 tomatoes, peeled, seeded, and cut into pieces*
> *2 teaspoons dried oregano*
> *2 teaspoons dried thyme*
> *2 bay leaves, broken*
> *Freshly ground black pepper*
> *1½ cups dry white wine*
> *6 unpeeled boiling potatoes, sliced ½ inch thick*
> *⅔ cup olives, such as unpitted Niçoise or Ponentine*
> *½ cup chopped parsley*
> *A cruet of fine olive oil*

Soak the cod in water in the refrigerator for two days; change the
water three times. When the cod has been desalted, remove the skin,
tail, bones, and dark spots. Tear the flesh into small pieces and
marinate it for 2 to 3 hours in a bowl with the olive oil.

In a heavy-bottomed pot, heat 2 or more tablespoons of the olive
oil used to marinate the fish. Add the cod and sauté it for 5 minutes,
stirring once. Add an additional tablespoon of olive oil, the garlic,
onions, bell peppers, tomatoes, oregano, thyme, bay leaves, pepper,
and wine. Simmer, covered, for 1½ hours. Add the sliced potatoes
and cook, covered, for 25 minutes more, or until done. Just before
serving, stir in the olives and parsley. Check the seasoning—it should
be quite peppery. Your guests should generously douse their servings
with fine olive oil.

Serves 8.

❧ BAKED ROCKFISH VERACRUZ STYLE ❧

from *The California Seafood Cookbook*

Huachinango á la Veracruzana is perhaps Mexico's most famous seafood dish. Around the Gulf of Mexico it is made with red snapper, but it can also be made with any local rockfish or firm-fleshed white fish. Instead of a whole fish, you may use two smaller fish, or 1½ to 2 pounds of fillets.

4- to 5-pound rockfish, dressed
Juice of 1 lemon or lime
Salt
1 medium yellow onion, julienned
1 bell pepper, julienned
1 tablespoon chopped garlic
2 medium tomatoes, peeled, seeded, and chopped
2 tablespoons olive oil
¼ cup Sicilian-style olives, pitted and chopped
1 tablespoon capers
1 or 2 pickled chili peppers

Preheat oven to 450° F. Sprinkle the fish with lemon or lime juice and salt, and set aside. Sauté the onion, bell pepper, garlic, and tomatoes in olive oil until just softened; do not brown. (If using canned tomatoes, add them after the onion and pepper are soft.) Simmer the sauce until most of the liquid has evaporated.

For whole fish: Place the fish in an oiled, deep baking pan. Pour the sauce over the fish and scatter the olives, capers, and chilis on top. Cover the pan with a tight-fitting lid or foil. Bake 10 minutes to the inch, or until the flesh flakes easily from the tail. Serve with the accumulated sauce from the pan.

For fillets: Cut into individual portions; place on squares of foil large enough to enclose them. Top each with sauce, olives, capers, and chilis, seal the edges tightly, and bake 8 to 10 minutes. The fish is done when the thickest part is about to lose its translucency.

Serves 4.

SALMON, STEELHEAD, OR TROUT
❉ WITH OLIVE BUTTER ❉

This method of cooking fish poaches it, since no liquid escapes from the foil in which the fish is tightly wrapped. If you are fortunate enough to have a freshly caught steelhead or trout, allow about 35 minutes cooking time for a 12-ounce fish. Small, farm-raised salmon have begun to appear at the fish markets, but are a good deal less satisfactory in flavor than a chunk of larger salmon fresh from its natural environment. A 2-pound piece of ocean-caught salmon will take about an hour to cook by this method.

Fish
Fine olive oil
Garlic

Tear off a piece of aluminum foil about 4 inches longer than the fish or piece of fish you are going to bake. Sprinkle olive oil over the fish and rub it on all sides. Pierce the flesh of the fish every few inches and insert slivers of garlic. Then wrap the fish, sealing it as well as you can on the top, so that no juices leak out or evaporate. Put the fish in a baking pan and bake at 300°F, 35 to 60 minutes (see above), depending on the size of the fish. Serve hot topped with olive butter.

Olive Butter

¼ pound Salona, Kalamata, or other ripe olives packed with vinegar
4 tablespoons sweet butter, room temperature
½ teaspoon dried oregano or 1 tablespoon chopped fresh sweet basil

Pit the olives and make a paste of them in a mortar. In a bowl, combine the olive paste, softened butter, and herbs. Serve over hot, boned fish (above) or sautéed chicken breasts.

Serves 3.

BARBECUED PORK
❀ IN OLIVE OIL MARINADE ❀

This barbecued pork is much like an unforgettable marinated pork I had in Tepic. Allow the meat at least 12 hours to absorb all the flavors. The pork should be sliced very thin, with the grain, and cooked until just done over good-quality charcoal.

3- to 4-pound pork loin, with or without the bone
1 cup olive oil
½ cup white wine vinegar
Tabasco sauce
Sugar
Cumin seeds, crushed to a fine powder in a mortar
Salt
Garlic cloves, crushed
1 large yellow onion, thinly sliced
Juniper berries
1 bunch cilantro, chopped

Slice the pork with the grain into thin pieces. In a high-sided bowl, layer the above ingredients as follows: Start with one layer of meat. Sprinkle it with a little olive oil, a little vinegar, a few dashes of Tabasco sauce, a sprinkling of sugar, a sprinkling of cumin seeds, a sprinkling of salt, a garlic clove, a few separated slices of onion, and a few juniper berries, crushed between your fingers. Repeat until all the meat is in the bowl; then cover with the remainder of the olive oil and vinegar. Marinate in the refrigerator. Remove a few hours before barbecuing to allow the meat to come to room temperature.

Barbecue until just done, on both sides, topping the second side with onions. Garnish with the onions and cilantro.

Serves 8.

MOROCCAN LAMB TAGINE
❄ WITH PRUNES AND OLIVES ❄

This typical Moroccan tagine, or stew, is easy to make and can be made any time of the year. (Tart apples from New Zealand are filling the northern hemisphere winter and spring apple void.) The sweetness of the prunes and honey is a nice contrast to the bitter olives.

3 pounds lamb shoulder, trimmed and cut into pieces
3 tablespoons olive oil
1 teaspoon salt
Pinch of crushed saffron threads
Good pinch of cayenne pepper
1 heaping teaspoon finely chopped fresh ginger
1/3 teaspoon cinnamon
1 yellow onion, half minced and half thinly sliced
2 cloves garlic, peeled and chopped
Water
3/4 cup dry-cured black olives
1/2 pound prunes, pitted and plumped in warm water
1 tablespoon toasted sesame seeds
1/2 tablespoon honey
1 bunch cilantro, chopped
2 tart apples, peeled and sliced
2 tablespoons unsalted butter

In an enameled casserole, sauté the lamb in the olive oil until browned. Add the salt, saffron, cayenne, ginger, cinnamon, minced onion, garlic, and water to cover. Stir; then cover the mixture and cook until tender, about 1 hour. Add the olives, prunes, sesame seeds, honey, cilantro, and sliced onion; simmer, covered, about 5 minutes.

Sauté slices of peeled apples in butter until soft. Serve the tagine over bulgur wheat or with couscous, garnished with the sautéed apple slices.

Serves 8.

❧ OLIVE-STUFFED LEG OF LAMB ❧

*T*here's nothing like a large leg of lamb for a handsome presentation at a special dinner. This Provençal recipe lends itself beautifully to a meat that is not overpowered by a flavorful marinade and stuffing. Oven-fried potatoes (page 72) make a nice complement when served surrounding the lamb on a bed of greens, such as watercress. If you have mastered barbecuing lamb, this recipe will be even better.

10 cloves garlic, peeled
Olive oil
⅓ pound cracked green olives, pitted
2 ounces sun-dried tomatoes, packed in olive oil
5- to 6-pound leg of lamb, boned
Bottle dry white wine
3 bay leaves, broken
1 yellow onion, sliced

Sauté the garlic cloves in 2 tablespoons of olive oil until golden brown. Chop the olives and tomatoes together very fine; then add the sautéed, whole garlic cloves. Stuff the lamb with the olive mixture and place it in a high-sided bowl. Pour 1 cup of olive oil and the wine (less ½ cup held in reserve for deglazing the pan) over the lamb and add the bay leaves and onion. Marinate the lamb in the refrigerator overnight, turning it once before bed and once in the morning.

Remove the meat from the refrigerator a few hours before cooking. Preheat the oven to 400°F. Place the meat in a baking pan and put it in the oven. Bake 15 minutes on one side; then turn and bake 15 minutes more. Turn down the oven to 350°F and continue baking for 50 minutes more, until the meat is cooked rare, a total of 1 hour and 20 minutes. For medium lamb, add 5 minutes per pound.

Skim most of the fat from the pan juices. Deglaze with the reserved ½ cup of wine, and reduce a little. Correct the seasoning and serve either spooned over the sliced lamb on a serving platter or in a sauce boat.

Slice the lamb with the grain and garnish with the stuffing.

Serves 10.

❀ *LAMB AND OLIVE BALLS* ❀

These piquant meatballs can be served plain with bulgur wheat or in sandwiches, or in a spicy tomato sauce over pasta.

3 slices bread, country-style white or whole wheat
2 pounds fairly lean ground lamb
¼ pound feta cheese, crumbled
1 cup Kalamata olives, pitted and chopped
1 egg, beaten
½ tablespoon cinnamon
½ teaspoon hot red pepper flakes
3 cloves garlic, crushed
1 bunch cilantro, chopped
3 tablespoons olive oil

Cut the crusts from the bread; soak the slices in water, wring them out, and crumble them. With your fingers, mix the lamb well with the bread, feta, olives, egg, cinnamon, hot pepper flakes, garlic, and cilantro. Form into 10 large meatballs.

In a heavy frying pan, cook the meatballs in the olive oil until crisp and brown on one side; then turn and brown the balls on all sides, no more than 10 minutes, over a fairly high heat. The meat should be rare.

Makes 10 meatballs.

LAMB SHANKS WITH FETA,
❧ RED BELL PEPPERS, AND OLIVES ❧

"This is what beef Stroganoff would have been if it were made by a Greek," said my friend after polishing off a helping. He had failed to see its subtlety: The sweetness of the red bell peppers provides a nice contrast to the saltiness of the feta and olives; the watercress adds a crisp lightness to an otherwise rich dish.

Purists will want to scorch and remove the skins from the peppers before sautéing them.

3 or 4 meaty lamb shanks, cut in half by the butcher
8 tablespoons olive oil
12 large cloves garlic
1 tablespoon dried oregano
Salt
Freshly ground black pepper
¼ pound creamy feta cheese
¼ pound cream cheese (preferably without gum)
Scant ¼ cup half-and-half
10 red bell peppers, washed and seeded
¾ cup dry red wine
1 small bunch fresh basil or other fresh Mediterranean herb, chopped
1 bunch watercress
24 Kalamata olives (approximately)

In a large, heavy skillet that has a lid, brown the lamb shanks on all sides in 4 tablespoons olive oil. Add 6 unpeeled garlic cloves, the oregano, and salt and pepper to taste. Cover and cook over a very low heat until the meat can be easily cut from the bones, about 2 hours.

Meanwhile, put the feta through a food mill, or blend in a blender or food processor until smooth. Cream together the feta, cream cheese, and half-and-half. Allow to stand at room temperature.

Cut the bell peppers into thin strips. In a skillet, cook the peppers in the remaining 4 tablespoons olive oil with the remaining 6 cloves garlic, crushed, a *little* salt, and pepper to taste, stirring from time to time until soft and shriveled, about 15 minutes.

Remove the meat and garlic cloves from the pan and cut the meat from its bones in large chunks. Pour off the fat from the pan and deglaze the remaining juices and remnants with wine. When the deglazed sauce has thickened sufficiently, remove from the heat and squeeze the whole garlic cloves from their peels into the pan. Add the fresh herbs and stir.

On each dinner plate, surround the meat with the red peppers on a bed of watercress. Off center, add a dollop of feta cheese mixture, and garnish with a few olives. Pour the deglazing sauce over all and serve with bulgur wheat, barley, or similar accompaniment.

Serves 4 to 6.

❧ *DAUBE OF BEEF* ❧

*T*here are probably as many different types of daubes as there are cooks who make them. The unifying characteristic of daubes, however, is that they use a large cut of meat (or a fowl) and braise it in wine. A beef daube can be made with a whole rump, or other cut, or with beef cut into pieces as for a stew. It should not be made from a tender, expensive cut, because the flavor benefits from the long, slow cooking that only humbler cuts of meat can withstand.

The following daube uses unmarinated chuck roast cut into large pieces. The *daubière*—a casserole dish named for the daube and made of stoneware, earthenware, or tinned copper—has here been replaced by a thick iron casserole (enameled or not), which can be left on the stove top for hours.

1 cup dried small white beans, rinsed
4 pounds chuck roast, cut into large pieces
¾ cup flour
Salt
9 tablespoons olive oil or more
2 cups dry white wine
4 cloves garlic, minced
1 yellow onion, thinly sliced
1 teaspoon dried thyme
2 bay leaves, broken in half
3 wide strips orange zest
Hot red pepper flakes
½ cup chopped fresh parsley, tightly packed
2 large carrots, sliced
10 cherry tomatoes, halved
2 handfuls fresh mushrooms
3 large unpeeled potatoes, thinly sliced
1 cup small French black olives, such as Niçoises
Grated orange zest
Chopped cilantro

Cook the small white beans in boiling water until tender, about 1¼ hours, and drain.

While they cook, dredge the beef in the flour mixed with 1 teaspoon salt. Brown in 4 tablespoons of the olive oil in a large casserole. Add the wine, garlic, onion, thyme, bay leaves, orange zest strips, and hot pepper flakes to taste. Cover and simmer about 1½ hours, or until the meat is half an hour from being done. Adjust the salt. Add the parsley, carrots, tomatoes, and cooked beans; cover and simmer half an hour more.

Meanwhile, in a large frying pan with a lid, sauté the mushrooms in 3 tablespoons of the olive oil and cook until done—that is, until they no longer absorb oil. Remove from the pan and set aside. Spread the sliced potatoes in the oiled pan and dribble the last 2 tablespoons of olive oil over them, salt lightly, and cover. Cook over low heat about 20 minutes, or until tender.

When the beef is done, add the mushrooms and olives to the daube and stir. Serve over the potatoes and garnish with grated orange zest and cilantro.

Serves 6.

❀ *BEEF TONGUE WITH PIQUANT SAUCE* ❀

I first had this informal dish while staying in Sarlat, in the south of France. The sauce is a delicious pickle, good with any boiled meat— ham, fowl, or, as here, tongue.

1 beef tongue
30 juniper berries, crushed
2 cinnamon sticks
3 bay leaves, broken
½ tablespoon salt

Rinse the tongue with cold water. Place it in a pot, cover it with water, and add the juniper berries, cinnamon sticks, bay leaves, and salt. Bring the liquid to a boil and simmer, covered, 2 to 3 hours, or until tender. Allow to cool before removing the skin. Slice and serve at room temperature with the following sauce.

Piquant Sauce

1 pound pearl onions, peeled
2 tomatoes, peeled and chopped
6 ounces tomato paste
7 tablespoons cider vinegar
1 cup raisins
½ cup water
1 cup Sicilian-style olives, pitted and chopped
3 tablespoons brown sugar
3 pinches cayenne pepper
¼ teaspoon cinnamon
Dried thyme to taste
1 cup walnuts, quartered and toasted

In boiling water, cook the onions 10 minutes. Drain. In a medium saucepan, heat the onions with the tomatoes, tomato paste, vinegar, raisins, water, olives, sugar, cayenne, cinnamon, and thyme. Stir and simmer, covered, until raisins are plump, about 5 minutes. Mix in the walnuts. Store in a covered jar in the refrigerator. The sauce is better after a few days and will keep about a week.

Serves 6.

❊ CARNE RELLENADA ❊

This succulent Spanish beef dish can be served cool (after being weighted in a deep dish in the refrigerator), or hot with its wonderful gravy. This is one of the most aromatic dishes I know, and one of the prettiest when sliced, with the yellow eggs and red peppers in a spiral.

3 eggs
Olive oil
2-pound flank steak, butterflied
3 cloves garlic, crushed
Salt
Freshly ground black pepper
½ pound Spanish- or Italian-style pork sausage, uncooked
2 red bell peppers, seeded and cut into strips
3 carrots, peeled and cut into thin strips
⅔ cup Spanish-style green olives, pitted and chopped
String
All-purpose flour
2 cups fresh beef stock or canned chicken stock
1 cup dry white wine
1 yellow onion, quartered
1 tomato, quartered
2 bay leaves, crumbled

Make a stiff omelet with the eggs and a little olive oil; remove it from the pan, cut the omelet into ½-inch-wide strips and set aside. Rub the butterflied steak with the garlic; then sprinkle it with salt and pepper. Remove the sausage from its casing and crumble it over the steak. Sauté the peppers and carrots in olive oil until slightly limp. Distribute them lengthwise over the steak along with the omelet strips. Sprinkle the olives evenly over the steak. Then roll the steak and tie securely with string in several places. Dredge the stuffed steak in flour and brown in olive oil in a heavy casserole. Add the stock, wine, onion, tomato, and bay leaves; cover and simmer over low heat for 1 to 1½ hours.

Transfer the meat to a cutting board. If it is served hot, cut and remove the strings and slice the beef roll. Strain the gravy into a sauce boat.

If it is served cold, put the rolled steak in a tightly fitting pan, remove the strings, and strain the gravy over the meat in the pan. Weight it and refrigerate for at least 2 hours.

Serves 8.

❧ *HAUTE TAMALES* ❧

Real, honest-to-goodness tamales have no place in an olive recipe book; but because tamale pie does (by virtue of its pitted, "black-ripe" olives) and since tamales are tastier and more fun than tamale pie (if ten times harder to make), I include a recipe for tamales with pitted California olives, and a little olive oil used in lieu of lard. The recipe was too good to pass up.

Try your neighborhood Mexican delicatessen, or a larger Mexican grocery, as a source of fresh masa (dried whole-kernel corn, ground with water into a paste) and corn husks. Once you become used to making tamales, the process goes quickly.

Begin by soaking the corn husks in water for at least an hour before you assemble the tamales.

Tamale Filling

2½ pounds chuck roast
1 cup all-purpose flour
1½ teaspoons salt
1 teaspoon dried oregano
½ cup olive oil
2 cups water
2 tablespoons chili powder
1 tablespoon cumin
5 cloves garlic, crushed
1 yellow onion, diced

Cut the meat from the bones into chunks and dredge in a mixture of the flour, ½ teaspoon of the salt, and ½ teaspoon of the oregano. Sauté the meat in the olive oil until browned on all sides. Drain the fat from the pan. Add the water, the remaining 1 teaspoon salt, the remaining ½ teaspoon oregano, chili powder, cumin, and garlic to the meat. Stir; cover and simmer for 1½ hours, stirring occasionally. Add the onion; stir and cook half an hour more, or until the meat can be broken apart easily. Let cool.

Masa

2 pounds fresh masa dough, without lard and unseasoned
1 tablespoon baking powder
1 cup olive oil
½ tablespoon salt
⅔ cup chicken stock (with fat if homemade)

In a bowl, combine the masa and the baking powder. Add the olive oil, salt, and chicken stock and blend well with a fork.

Dried corn husks, soaked
1 can pitted California "black-ripe" olives

To assemble: Rinse the corn husks and shake dry. Using the large ones, place two pieces of husk side by side, wide end at the top for one and narrow end at the top for the other, overlapping about ⅓ inch. The two will roughly form a square. Spread 3 tablespoons of the masa dough over the middle third (as if the square were divided into three horizontal strips), leaving about ½ inch of husk uncovered on one vertical side as well as the top and bottom thirds. Make sure the seam of the two husks is well covered with masa.

Spread one large spoonful of the filling over the masa, breaking the meat with your fingers and not sparing the gravy. The meat and gravy should come to within ⅓ inch of the edge of the masa on all sides. Place 2 or 3 pitted olives on the filling. Fold the side with the masa spread to the edge, then fold the other side, overlapping at least ½ inch. Gather the ends together, making a pouch for the tamale. With strips torn from corn husks, tie the ends of the tamale tightly. Steam 45 minutes in a steamer.

Makes 15 tamales.

CHICKEN WITH FETA
❧ AND KALAMATA OLIVES ❧

This savory Greek-style dish was adapted from a recipe by Ann Walker, a Bay Area caterer known for her robust cuisine.

½ cup flour
1 teaspoon salt
½ teaspoon freshly ground black pepper
½ teaspoon dried marjoram
½ teaspoon dried oregano
1 frying chicken, cut up
¼ cup olive oil
1 cup chicken stock (made from the back, neck, and so on)
Juice of 1 lemon
2 cups fresh or canned tomatoes, peeled, chopped, and with their juice
25 Kalamata olives
1 large garlic clove, crushed
½ pound feta, crumbled
2 tablespoons fresh basil or other fresh Mediterranean herb

Combine the flour, salt, pepper, marjoram, and oregano. Dredge the chicken in the mixture, reserving the remainder. In a large skillet, heat the olive oil until a haze forms above it, and brown the chicken on all sides. (Cook the legs and thighs a little longer than the breasts and wings.) Remove the chicken to a heavy, lidded casserole.

Drain off all but 2 tablespoons of the fat from the skillet. Stir 3 tablespoons of the reserved flour mixture into the fat in the skillet, and gradually add the stock and lemon juice over a low heat, stirring with a wire whisk until thick and smooth; then add the tomatoes, olives, and garlic. Stir and pour over the chicken. Cover and cook over medium heat until the breasts are just done, about 30 minutes. Scatter the feta on top, cover, and cook a few minutes until the cheese melts. Garnish with the fresh herbs.

Serves 4.

CHICKEN WITH LEEKS, TARRAGON,
❋ ORANGE, AND OLIVES ❋

This seasonal dish relies on the availability of fresh tarragon—the character of dried tarragon is quite different. This is a delicate, refreshing dish, good for a special luncheon.

Frying chicken, cut up
⅓ cup flour
Salt
Freshly ground black pepper
3 leeks
¼ cup olive oil
4 cups chicken stock
1 orange, very thinly sliced
2 teaspoons chopped fresh tarragon
1 cup tiny black brine-cured olives, such as Niçoise or Ponentine

In a paper bag, shake the chicken parts in the flour, 1 teaspoon salt, and pepper to taste. Halve the leeks lengthwise, wash thoroughly, then halve crosswise. In a large enamel casserole with a lid, heat the olive oil to very hot, and place the leeks side by side, cut-side down; cook until browned. Remove from the casserole. Brown the chicken pieces, cooking the breasts less than the legs and thighs, and drain on paper towels. Drain the chicken fat and oil from the casserole and add the chicken, leeks, chicken stock, orange slices, and tarragon. Cover and simmer for approximately half an hour, or until the breasts are just tender. Remove the chicken and place on a serving dish in a warm oven. Then bring the liquid and the leeks to a boil and cook the leeks, uncovered, a few minutes until tender. Remove the leeks and arrange them attractively around the chicken and replace it in the oven. Now cook the liquid at a rapid boil until it is reduced and thickened. Add the olives, adjust the salt, and pour the sauce over the chicken. Serve with bulgur wheat.

Serves 4.

MOROCCAN CHICKEN WITH
❧ CRACKED GREEN OLIVES ❧

This chicken, served whole and tinted with saffron, makes a handsome presentation. Carve the chicken at the table and serve with plenty of lemony gravy.

2 cups (about ¾ pound) cracked green olives
4 tablespoons olive oil
3½- to 4-pound chicken, whole
Salt
Freshly ground black pepper
1 teaspoon minced fresh ginger
3 cloves garlic, minced
1 tablespoon cumin seeds, ground
1 large pinch saffron threads, crushed with your fingers
2½ cups light chicken stock
Lettuce, red leaf or curly endive
¼ cup fresh lemon juice
Zest of 1 lemon, grated

Pit the olives; then boil them for 15 minutes in water. Drain and boil the olives again for 15 minutes (this process makes them less bitter, but dulls their bright green color). Set aside. In a large casserole, with 2 tablespoons of the olive oil, brown the chicken well on all sides. Remove from the casserole and drain the fat from the pot. Salt and pepper the chicken to taste. In the same pot, sauté the ginger, garlic, cumin, and saffron in the remaining 2 tablespoons olive oil for 1 minute; then add the chicken stock and stir. Return the chicken to the pot, cover, and cook for 12 minutes on one side. Then turn the chicken and cook, covered, for 12 minutes on the other side. The chicken should be just done. If it is not, cook for a few more minutes, and test again.

Remove the chicken from the liquid and put it in a serving dish on a bed of lettuce. Add the lemon juice to the liquid in the casserole and reduce until the sauce is slightly thickened. Add the olives just long enough to heat them. With a slotted spoon, distribute the olives over the chicken. Serve the gravy in a gravy boat at the table. Garnish the chicken with the grated lemon zest.

Serves 4.

❧ DUCK WITH WINE AND OLIVES ❧

The following recipe provides a number of flavors and textures that balance each other beautifully: the rich duck cut by the tart lemon zest and complemented by the sweet wine; the cooked onions and the firm olives enveloped by garlic and thyme, the flavors of Provence.

4- to 5-pound duck
Salt
2 yellow onions
1¼ cups sweet, fruity wine, such as a Spätlese or Muscat
1 teaspoon dried thyme
6 small cloves garlic, peeled
Lemon zest
Freshly ground black pepper
¼ pound small olives, such as Niçoise, picholine, or Ponentine
Watercress

Preheat the oven to 425°F. Remove the wing tips from the bird, as they will burn upon roasting. Remove the excess fat from the neck end of the bird and from the body cavity. Rub the bird inside and out with salt. Cut up one of the onions and put it in the body cavity. Truss the bird as you would a chicken to keep the wings and legs close to the body. With a fork, poke the bird all over at close intervals so the fat can escape as the bird cooks.

Put the duck breast-side up in a roasting pan and bake until lightly browned, about 20 minutes. Turn the bird over; reduce heat to 350°F and bake for 1½ hours more. (If you prefer a pink meat, cook the duck a total of 1¼ hours, or until the juices run a pale pink when the bird is pricked near the thigh joint. The less-done bird has better flavor and moisture, but can be a little tougher.)

Make a stock while the duck bakes. In a saucepan, cover the gizzard, neck, heart, and wing tips with water; add the remaining onion, coarsely chopped, and simmer, covered, for an hour or so. Remove the duck pieces and onion and reduce the liquid to ¼ cup. Add the wine, thyme, garlic, a few strips of lemon zest, a pinch of salt, pepper, and the duck juices (removed with a bulb baster from beneath the fat in the duck pan). Over high heat, reduce the liquid by half. Add the olives.

Cut the duck into serving pieces and arrange on a bed of watercress. Pour the sauce over all.

Serves 3.

❧ RABBIT STEW WITH OLIVES ❧

*P*eter, Bugs, and Easter have put a damper on rabbit consumption in America, despite the fine qualities of the dark, juicy meat of bunnies. If good, fresh, and inexpensive rabbit is available to you, the following recipe makes a tasty, hearty stew.

Serve with a robust bread to sop up the thick, plentiful gravy this recipe makes.

6 pounds rabbit, preferably stewing rabbit
¾ cup flour
Salt
Freshly ground black pepper
⅓ cup olive oil
1½ cups medium chicken stock
Bottle dry white wine
4 cloves garlic, minced
3 bay leaves, broken
1 cup finely chopped parsley
1 teaspoon dried thyme
4 tablespoons tomato paste
¼ pound hot coppa or ham, cut into strips
½ pound dry-cured black olives
Grated zest of 1 lemon

Dredge the pieces of rabbit in flour seasoned with salt and a good dose of finely ground pepper. Brown in the olive oil in a large, heavy casserole, draining the pieces on paper towels as they are done and adding more olive oil if necessary. When all the pieces are browned, pour out the fat and wipe the pot clean. Replace the meat and add chicken stock, wine, garlic, bay leaves, ½ cup of the parsley, the thyme, and tomato paste. Stir; cover and simmer until the rabbit is tender, up to 1¾ hours, depending on the age of the animal. Add the coppa and olives and cook 5 minutes more.

Serve garnished with the remaining ½ cup parsley and the grated lemon zest.

Serves 8.

Desserts

❧ *IL CASTAGNACCIO* ❧
Chestnut Flour Cake

*T*he recipe for this interesting Tuscan cake dates from the Middle Ages and is still much loved by some Tuscans. After serving the olive oil and rosemary flavored castagnaccio at a dinner party, I was told by a polite guest, "I like it because it's more resistable than the average dessert." I can't think of anything that would improve it, save using wheat instead of chestnut flour, eggs instead of milk, and so on. But then it wouldn't be castagnaccio. It is a heavy, earthy-looking cake; you can count on most people not coming back for seconds. For tradition's sake, I include the recipe; try to get very fresh chestnut flour for the best results.

2 cups sifted chestnut flour
1 tablespoon sugar
Pinch of salt
3 tablespoons pine nuts
Scant 2 cups milk
3 tablespoons olive oil
⅓ cup raisins, soaked in milk
1 teaspoon chopped fresh rosemary

Preheat the oven to 425°F. In a mixing bowl, mix the chestnut flour with the sugar, salt, and 2 tablespoons of the pine nuts. Slowly add the milk, stirring constantly so that lumps do not form. The batter will be quite fluid. Mix in 2 tablespoons of the olive oil. Drain the raisins and toss them with a little chestnut flour so that each is coated with flour; then mix them into the batter.

Grease a round, 9-inch cake or pie pan with olive oil and pour in the batter. Sprinkle with the rosemary and the remaining pine nuts and olive oil. Bake in a 425°F oven until the top is lightly browned, about half an hour. Remove from the oven and let rest 10 minutes. Serve hot or at room temperature, cut in wedges.

Serves 8 to 10.

❧ SOUFFLÉ FRITTERS ❧

*T*he mixture for making these puffs is nothing more than a *chou* pastry, the kind used in making éclairs and profiteroles. Here, however, it is fried in olive oil and served plain, simply dusted with powdered sugar while still hot.

¾ cup water
¼ cup unsalted butter, cut into pieces
1 teaspoon sugar
Scant 1 teaspoon grated lemon zest
Pinch of salt
¾ cup flour, sifted
3 eggs
Olive oil for frying
Powdered sugar

Combine the water, butter, sugar, lemon zest, and salt in a saucepan and heat, stirring, until the butter has just melted. Remove from the flame. Add the flour all at once and, slowly at first, blend in the flour with a wooden spoon. Return to a medium-low flame and, stirring constantly, dry the flour paste for a few minutes until it comes away from the pan and adheres in a ball. Remove from the flame. Make a well in the paste and add 1 of the eggs. Stir vigorously until it is thoroughly incorporated. Make another well. Add another egg, stir, and repeat the process with the last egg.

Heat about ¾ inch olive oil in a frying pan (even an omelet-size pan will do for frying four puffs at a time). Spoon a generous dollop into the pan for each puff, keeping in mind that the puffs will expand to about twice their original size and must not run into one another. Keep the olive oil bubbling, being careful that it doesn't reach the smoking stage. Cook the puffs about 1¾ minutes on each side, turning with a slotted spoon, until they are a rich brown. Break one puff open to make sure the dough is cooked through. Serve hot with plenty of powdered sugar sifted over them.

Makes 20 fritters.

❧ CITRUS AND ALMOND CAKE ❧

This cake, though heavy, is refreshing in its tartness. Be sure to use a flavorful olive oil—the fine, more delicate oils are overpowered by the citrus rinds. It can be served with crème fraîche or lightly sweetened whipped cream, but it is delicious and moist by itself.

2 small navel oranges
1 lemon
6 ounces almonds
4 eggs
½ teaspoon salt
1½ cups sugar
1 cup all-purpose flour
3 teaspoons baking powder
⅔ cup olive oil

In a small saucepan, cover the oranges and lemon with water and simmer for half an hour. Drain and let cool. Cut off the stem ends, and cut the fruit in half; scoop out the pulp and seeds of the lemon and discard; then chop the oranges (with rind) and lemon rind very fine. Squeeze out as much liquid as you can in a strainer.

Chop the almonds in a blender or food processor until almost as fine as crumbs.

In a bowl, beat the eggs and salt until very thick and light. Then gradually add the sugar while continuing to beat.

Mix the flour and baking powder and add to the egg mixture until blended. Mix in the fruit, nuts, and olive oil, being careful not to overmix.

Turn the batter into an oiled, 9-inch springform pan. Bake at 350°F for 1 hour, or until a knife inserted into the center comes out clean.

Serves 8.

❄ *MELOMACARONA* ❄
Greek Cookies Made with Olive Oil

*I*n the Greek countryside these cookies are made by the kilo, and one of the ingredients is wood ash. This version might be a little less earthy.

Some people prefer these cookies without honey syrup and walnuts, and I must say they are a more subtle treat without them. Cut the amount of ground walnuts to ½ cup if you plan on serving them plain.

½ cup sweet butter, room temperature
¼ cup olive oil
¼ cup sugar
1 egg
½ teaspoon cinnamon
Juice of 1 orange
Zest of 1 orange, finely grated
2¼ cups unsifted all-purpose flour
1½ teaspoons baking powder
Pinch of salt
1 cup finely ground walnuts, lightly toasted
Honey syrup (⅛ cup water heated with ½ cup honey)

Preheat the oven to 375°F. Beat the butter until it is creamy. Add the olive oil and sugar and beat until fluffy. Add the egg, cinnamon, orange juice, and orange zest, and stir well. In a separate bowl, mix together the flour, baking powder, and salt; then add to the butter mixture a little at a time. At first the mixture can be stirred, but when it thickens, turn it onto a floured bread board and knead in the rest of the flour mixture. Then knead ½ cup of the walnuts into the dough.

Form the dough into a long roll about 1½ inches in diameter. Flatten it slightly and cut it into ¼-inch slices. Arrange the oval-shaped cookies on a baking sheet lightly greased with olive oil. Bake in a preheated 375°F oven until the cookies are a delicate golden brown, about 20 to 25 minutes.

To serve, dip the cookies in the warm honey syrup and roll them in the remaining walnuts.

Makes approximately 2 dozen cookies.

The Olive in America

pim·o′la (pĭm·ō′la) *n.*
An olive stuffed with pimiento.

Webster's *New International
Dictionary of the
English Language*

What's in a name?

Shakespeare

*A*mericans have held up well under the strain of isolation. Being out here in the middle of the Atlantic (or Pacific, depending on your point of orientation) away from our cultural roots, we have developed our own music, our own architecture, our own sports, and, of course, our own olive, the pitted "black-ripe."

In keeping with much of America's eating tradition, the American olive is a subtle-tasting item—production-line all the way. When an American opens a can of super colossals, each one will be the spitting image of the fellow on the label. No deviations, no blemishes—perfect. Americans like their "ripe" olives crisp, shiny, and black, in a mild, uncloudy brine. And the bigger, the better.

Although it is the Chevy Bel Air of olives, even the "black-ripe" olive is too exotic for some Americans. The California olive industry, which produces over 99 percent of the nation's olives, is having a rough time selling olives in many of the country's major markets— especially to southerners. For example, Atlanta-area consumers buy a paltry 0.46 percent of the nation's olives, compared with 16.5 percent for the Los Angeles area, and 9 percent for New York.

Nevertheless, over the past ten years per capita consumption of canned "black-ripe" olives has risen from 0.44 pound in 1971/72 to 0.55 pound in 1980/81. That rise, and an increased U.S. population, mean the American olive industry is in pretty good shape. Except in a few gourmet ghettos, the "black-ripe" is meeting with little competition for the patronage of American olive eaters.

THE OLIVE CROP

The olive is one of the few crops in America to be grown principally on farms of moderate size. Even the few olive acreages owned by large corporations are small when compared with some of the super colossal landholdings typical of crops that are highly mechanized. Because olives are hand-harvested, and require pruning and tilling, there are few cost benefits to be realized by gargantuan-scale farming. Many of the olive orchards, particularly in northern California, are between fifteen and twenty acres.

The ideal climate for the olive is one with a long, hot summer; winter chilling sufficient to set fruit (minimum temperatures of no less than 12°F); and no late spring frosts to kill the blossoms. Olives are concentrated in eight major olive-growing counties in two areas of California. Tulare and Kern counties in the Central Valley grow almost half the olives in the state; Tehama and Butte counties in the Sacramento Valley in the north grow about a fifth. (There is a small pocket of olive acreage outside Phoenix which produces about 1 percent of what California produces.)

Five commercially important varieties of olive are grown in California: Manzanillo (pronounced man-zan-ill-o by those who grow them), Sevillano (sev-ill-ahn-o), Mission, Ascolano, and Barouni, listed in descending order of crop size. The Manzanillo is the most abundantly grown because of its fairly large size, its ease of processing and resistance to bruising, and because it ripens early, enabling growers to harvest before the early frosts. The Sevillano, sometimes called the queen olive, is next in importance because it is the largest olive and therefore fetches the highest prices at the grocery store. It was the Sevillano that confused the size designations on labels. The Mission was the first olive on the scene, with sizes designated as small, medium, large, and extra large. Thus the Sevillano, the smallest of which is larger than the largest Mission, has designations giant, jumbo, colossal, super colossal, and special super colossal. Because the Sevillano has a low oil content, it cannot be used for oil extraction as can Missions, Manzanillos, and even Ascolanos. The largest production of the Sevillano is in Tehama County, north of Sacramento.

Growing in nearby Butte County are most of the state's remaining Mission olives, the first olive to be cultivated in America. It is the smallest of the California olives, and was originally planted for its high oil yield; but because California olive oil producers find it hard to compete with importers, the Mission olive trees are being top-grafted to larger table olive varieties (Manzanillos and Sevillanos) and the Mission is being slowly replaced. However, the Mission remains the olive of choice for the state's cold-pressed oil makers.

The Italian Ascolano produces a large olive that is oilier than the Sevillano; but it bruises easily, so little acreage is planted. Still fewer acres are devoted to the Tunisian Barouni, which is a small olive shipped fresh to the East Coast for marketing to home-cured olive makers.

A DAY IN THE LIFE OF AN OLIVE

Let us follow a typical olive from a typical orchard near Corning—in the Sacramento Valley within view of Mt. Shasta—as the olive makes its way from the tree to the can as a "black-ripe" pitted olive. It grows on a handsome forty-year-old Manzanillo tree that has been pruned low and spreading to facilitate harvest. The harvest begins in early October, when the olives have achieved a good size and are a bright chartreuse, just right to be made into "black-ripe" olives. The pickers are migrant farm workers, few of whom speak any English. They stand on fifteen-foot, three-legged aluminum ladders, their hands partially taped for protection, and pick the olives in a milking fashion with both hands. Strapped to their bellies are buckets. Our hard olive drops into one of them. The pickers dump their buckets into wooden boxes that are stacked in the orchard rows, to be collected later by a pickup truck and taken to a receiving station where the olives are sized and graded under the supervision of the U.S. Department of Agriculture.

Once our olive and its peers have passed muster, they are trucked to one of the state's remaining handful of processor/canners. Though there are some cooperatives, such as Lindsay in the San Joaquin Valley, this one is an independent—Bell Carter, in the town of Corning. Our olive is dumped into a nine-ton-capacity fiberglass tub and soaked in a series of frothy lye baths that bubble as air is pumped through the liquid. The lye baths leach the bitter glucosides from the olives; the air "oxygenates" the olives, turning them from bright green to solid black. Minute quantities of ferrous gluconate, the same substance used in iron tablets, fix the color. The olives are then soaked in a mild salt brine (9.5 on the salometer), after which they go to the pitter—a fantastic machine that performs a thousand pittectomies a minute.

After it has been pitted, our olive is ready for inspection, and passes before a hundred eyes, namely, those of the women minding the conveyor belt. Being an acceptable olive, into a can it goes, with new brine and fellow Mammoth (i.e., medium-size) "black-ripes." The can is sealed and shoved into an enormous pressure cooker, which sterilizes the can and its contents at 262°F under careful supervision.

Our olive now waits to be freed from its container and to be waved about on the fingertip of a happy five-year-old (the true, 100-percent authentic way of serving California olives) and eaten.

PROCESSING OLIVES

Not all American-grown olives become "black-ripes," nor do all American ripe-olive eaters restrict themselves to the understated canned version. A small percentage of the California olive crop fits into the "other" category—olives crushed for olive oil, sold fresh, canned as green-ripe (not blackened with oxygen), or processed in brine for sale to specialty shops and delicatessens. The exact proportion varies greatly from year to year with the amount and size of olives harvested.

Anywhere from one twentieth to one sixth of the total California olive harvest is processed green, depending on the size of the harvest. The salty, acidic greens are known as "Spanish style," and for good reason: Most of the green olives eaten in America are imported from the world's largest olive producer, Spain, and taste just about the same as the domestic version.

The only stuffed olive manufactured in any quantity in California is the pimiento-stuffed olive. (Processes for other stuffings—onions and nuts, for example—have yet to be mechanized. Hand-stuffed olives are all from Spain, where labor expenses are much lower.) The people who market olives have come up with a process comparable to that for making the maraschino cherry (a fruit whose color and flavor are extracted and to which a new color and flavor are added). What starts out as a pimiento is ground up (seeds and all) and retexturized with carrageenan, then exuded from a machine in an endless, bright red ribbon with exactly the same circumference as a Spanish-style olive's pithole. The same process is used by Spanish processors near Seville.

A few small outfits in California are marketing tasty, mildly lye-cured olives, bottled in a number of herb- and spice-flavored brines. Such olives are finding a market in specialty stores at high prices. Similar effects can be achieved with less expensive, canned "tree-ripened" olives to which garlic, red peppers, or spices are added at home. ("Tree-ripened" olives are picked after they have begun to turn red; they are not oxygenated, and are therefore not black as are the "black-ripe" olives.)

Fine California-made European-style olives are turning up in deli cases on both coasts. One producer of such olives is West Coast Products in Orland, a town near Corning in the Sacramento Valley. Using local olives, which the company buys from the harvesters, West Coast makes a meaty oil-cured olive from dead-ripe Missions picked in March; a Greek-style olive from ripe olives picked between January and March; and Sicilian-style and cracked green olives from fall-harvested Sevillanos fermented in brine for up to six months. All the olives compare well with their imported counterparts, surpassing some of them because they have traveled less distance and have not suffered at the hands of shippers and warehousers.

CALIFORNIA OLIVE OIL

West Coast and a few other California companies also produce olive oil, and are competing for the rapidly growing virgin olive oil markets—people who care about good food and like the flavor of olive oil, and those who frequent health food stores. The producers are recovering from the nosedive the industry took in the 1940s, when inexpensive imported oils ran most California olive oil makers out of business.

West Coast sells varying qualities of oil, depending on when the pressing is done, and sells it to middlemen who either blend it or sell it under their own labels. It is shipped in barrels to Zabar's Delicatessen in New York City, where it is drawn from a spigot and sold to upscale cooks.

California's virgin oil makers label their products variously: Some print the year of the crop, some the variety of olive; some give nutritional information and boast the absence of preservatives. All use the designation they consider the most impressive or comprehensible—"virgin," "cold-pressed," and so on. Unfortunately, no legislation has yet been passed beyond that of making a distinction between refined oil and unrefined virgin oil. Domestic oil producers can use the designation "extra virgin" regardless of the quality of their product, as long as it is not refined. Many in the California olive oil industry would like to see a change. They sell a well-made product and would like that reflected by a credible label.

So far, even the best California oils do not compare with the finest imports. The Mission olive, the best of the California olives for oil production, is not comparable in quality to the smaller oil olives grown in some colder climates around the Mediterranean. And many California companies do not restrict themselves to pressing only Missions. Some San Joaquin Valley oil producers use the culls from canneries whose chief olives are Manzanillos. Often, the olives are pressed when they have achieved their highest oil content, but are past their peak of quality. And most of the domestic oil presses press the whole olive, pit and all, imparting a woody flavor to the oil. Still, oil preferences are largely a matter of individual taste, and some fine cooks find certain California oils quite to their liking.

With the virgin oil market expanding and the American consumer becoming more interested in good cuisine, it would seem likely that all facets of the American olive industry will be expanding in the near future.

HISTORY

If the future of the olive in America could be described as bright, its past might be described as curious: The mission fathers introduced it; land developers hawked it; the University of California perfected it; and fate conspired against its success. It was a pioneer crop in a pioneer state.

Early records. It is unlikely that the olive was brought to the missions at their founding in 1769, although the secular head in charge of establishing the settlements, José de Galvez, was far-seeing and had arranged for the shipment of seeds and grain for planting. The manifest of the ship that first brought supplies makes no mention, however, of olive seedlings. Instead, it is more likely that the olives — a variety appropriately dubbed "Mission"—were brought some time later, probably around 1785, for the purpose of oil production. Artisans arrived at the missions around 1795, and it was probably then that the first mills and screw presses were built.

The earliest record of oil production was by Father Lasuen, who wrote in 1803 that some missions had begun harvesting olives and that Mission San Diego had produced some very good oil. In 1834, the missions were secularized and their orchards largely neglected, although there are records that at some ranches, the Peralta Ranch, for example, Indians still tended the trees. Decades later, cuttings were made from the trees at the San Diego and San Jose missions by enterprising California nurserymen.

The land boom. For a number of reasons, not the least being that "not one American in ten thousand had tasted olive oil" (*Sacramento Bee*, 1896), and therefore little market existed for it, olives were slow to gain acceptance after the missionaries left. An 1855 agricultural report listed only 503 commercial olive trees in the state. An oil mill at Camulos in Ventura County, built in 1871, was probably the first to be constructed outside the missions. Slowly, however, a market for

Mission San Diego de Alacalá.

olives did start growing with the newly arrived non-Spanish immigrants. In fact, the olive became one of those exotic crops offered out-of-staters as an excellent investment, made all the more enticing because it could not be grown in the East, Midwest, or South.

By 1876, the number of trees had increased to 5603. By then, many energetic young pioneer growers had decided to grow olives, and there were many reports of successful plantings to encourage them. Frank and Warren Kimball were the first to plant on a large scale. In 1885, they planted a large farm in National City with cuttings taken from the old trees at nearby Mission San Diego. The Quito Ranch near Los Gatos in Santa Clara County had its own mill and, at eighty acres, the largest planting in central California. There, a few of the olives were pickled, but most were crushed for oil. (At Quito, black scale, even now one of the state's major olive pests, was treated with whale oil soap.) By 1885, hopeful speculators had planted olive orchards in Livermore (with transplants from Mission San Jose), Riverside, South Pasadena, Los Angeles, Berkeley, Sacramento, Auburn, Napa, and Solano. The 1896 U.S. Department of Agriculture *Yearbook* stated that in 1894 over 400,000 olive trees had been ordered from nurseries in Pomona alone. It is probable that a good portion of those trees were never planted; still, the orders show that many people considered the olive a potential gold mine. By the late 1800s, the olive had become a farming fad. Unfortunately, the oil produced by the new olive farmers could not compete with the much cheaper European imports (an unfortunate circumstance that occurred again in the 1940s), and the outlook was not bright for a better market opening up at the beginning of the twentieth century.

In the early 1890s, Warren Woodson bought 50,000 acres of railroad-land-grant property near Corning, in the hot northern Sacramento Valley. He planted apricots, peaches, plums, cherries, and Mission olives. Then, advertising in church circulars and at the World's Columbian Exposition, held in Chicago in 1893, he attracted midwesterners and easterners who wanted to become farmers in California, the land of opportunity. (The town of Corning still has an unusually high number of churches per capita, thanks to the ministers and deacons of various denominations attracted by Woodson's church advertisements.) During the first few years, many of the plantings were wiped out by high temperatures and drought. The hearty olive trees, however, survived. So Woodson quickly bought all the olive trees he could find, including a batch of imported trees from Spain, and replaced the more delicate crops with olives. (He planted the imported trees reluctantly, later to discover that they produced a larger, albeit less oily, fruit than the Missions: They were Sevillanos, the variety that produces today's super colossals.) The new residents of Corning, the first olive-growing community in the state, found themselves farmers of one crop, the olive.

Meanwhile, in the Central Valley, near a town called Lindsay in Tulare County, the olive was catching on, but in a more offhand way. There, the new crop was citrus. Olive trees served merely as ornamentals or windbreaks for the citrus trees. The early olive trees in Tulare were Missions, and thrived among the well-irrigated citrus. In some hill plantings the citrus did so poorly that they were replaced with olives, which would endure in poor, calcareous soils. It was soon well known that olives thrived around Lindsay; but because there was no sure market for olive oil or brine-cured olives during the '90s, the farmers of Lindsay continued to plant citrus and other fruit crops. When a method for canning was perfected more than a decade after the turn of the century, and a popular new "black-ripe" olive had been developed to assure a market for the California cured olive, Lindsay and the surrounding area became the most heavily planted olive region in the state.

The "black-ripe" olive. In 1892, an Illinois woman named Freda Ehmann came to California on the advice of her son, who owned an olive orchard about fifty miles from the Corning Colony in Oroville. When the Ehmanns' trees first bore fruit in 1897, Mrs. Ehmann contacted Professor Eugene Hilgard, an agriculturist at the University of California at Berkeley, requesting a recipe for processing ripe olives. Using Professor Hilgard's method for lye curing, she pickled olives on the back porch of her daughter's home in Oakland. It was that mild-flavored "ripe" olive that was to become the olive of California, and of America. Mrs. Ehmann's aggressive selling of the product, in barrels and glass jars, first to grocers in Oakland and then across the country to the East Coast and even up to the Klondike, enabled her to form a company in Oroville in 1898. Her two-ounce jars were served at dinners aboard the luxurious Southern Pacific and Santa Fe railroads; and fancy hotels served dishes featuring the much sought-after delicacy.

Almost a decade after Professor Hilgard divulged his recipe to Mrs. Ehmann, his colleague, Professor Frederic Bioletti, perfected a process for canning olives in tin. At that time, the olive industry had grown to over two million trees in the state, and farmers desperately needed a safe canning method to enable them to better control supply. Before the tin canning method had come into wide use, however, disaster hit. In 1919 and 1920, thirty-five people in the East and Midwest died of botulism after eating black olives packed in inadequately sterilized glass jars. The lye curing produced an alkaline brine that was a perfect medium for botulinus bacteria; it was almost inevitable that sooner or later a mistake would be made. After the tragic setback, the new high-temperature, high-pressure method of retorting slowly gained the trust of consumers, and the olive industry began rolling again.

The Berkeley group. California land development occurred in waves. The wave that followed Woodson's subdivision in Corning swept the state after 1910, when land was still cheap enough for developers to sell inexpensively to city dwellers who wanted to get rich quick or provide for their old age. During that time, an olive boom was rekindled in Oroville and attracted an unlikely group. Eight members of the faculty at the University of Nevada, and nine from the University of California at Berkeley—all graduates of Berkeley and from an assortment of disciplines—decided to invest together in agricultural land. For a minimal investment and the sweat of their own brows, they hoped to gain financial security for their families. Dr. Herbert Hill, an English professor, scouted the state for an appropriate crop. He considered oranges, figs, nuts, grapes, and peaches. He decided on olives; they live forever. (Perhaps their classical history had some appeal for the professor as well.) The two engineers in the group, Dr. Charles Hyde and Dr. Bernard Etcheverry, devised a hexagonal planting design that would facilitate cultivation. In 1914, Dr. Hill, Dr. Ralph Minor, a physicist, and Drs. Young and Frandsen began blasting holes in rock and setting out little trees. In ten years, the trees were self-supporting. Some of the professors came to Oroville in the summertime to work, and camped out in the orchards with their families. Professor Vaughn built a knotty pine cottage, to which he retired years later. During only two years did the venture show a loss, and that was during the Great Depression.

The growth of the industry—and the pitter. At the same time the Berkeley group was planting in Oroville, the olive industry was establishing itself in Lindsay. After Professor Bioletti's canning procedure became well known, a small group of Lindsay growers decided to start their own cooperative processing plant. It opened in 1916, with a capacity of 125 tons per year. In 1919 it was joined by a second plant, and Lindsay was well on its way to being the olive capital of the country—Lindsay, "a nice town, a great olive," as the highway billboard says.

All that was left to perfect—to eliminate that troubling moment when the olive eater must get the pit from mouth to pocket—was a good pitting machine. The California olive is one of the few olives marketed without the pit; and it was Herbert Kagley, a young mechanic, who devised the first mechanical olive pitter. In 1933, at the suggestion of the plant manager of Lindsay Ripe Olive Company, Kagley designed and built the prototype for the pitter. The machine was perfected by two engineers, Drake and Alberti; and it was later pitted, so to speak, against the Ashlock pitter. But H. Kagley made the one and only original, the machine that thrust the olive into twentieth-century high society and made the martini possible.

NOBLEST OLIVE OF THEM ALL*
by L. R. Shannon

The olive's noblest function is, of course, to keep a lemon twist out of your martini.

How a scrap of lemon skin ever replaced the olive in the first place is one of the great postwar mysteries. It deeply offends gin and bullies vermouth, which properly feels inferior anyway.

A respectable martini olive is firm, green and pitless. It is neither so large as to displace a significant quantity of liquid, nor so small as to be mistaken for a health code violation. Never does it contain a pimiento; a drink must not stare back.

What you do not do, however, is actually eat the olive. The martini consumed, the olive remains as a presence, a reminder, even a souvenir, although it is considered bad form to put it in your pocket as you might a swizzle stick or matchbook.

Each martini merits a fresh olive. The used ones can be thrown to the urchins pressing their noses against the window.

One or two uneaten olives are correct before a meal. Three frequently lead to equally uneaten meals. Four? You might as well start with a lemon twist and be done with it.

* Reprinted from *The New York Times*, January 9, 1980. Living Section.
© 1980 by The New York Times Company. Reprinted by permission.

The Botanical and Horticultural Olive

It did not knock me for a loop
To learn the olive is a drupe.
So is the peach, or nectarine,
So is the purple plum,
and green.
But olives are the prototype,
The drupiest drupes,
both green and ripe.
So, like the botanist who knows
That the apple is a rose,
Just think,
"The olive is a drupe,"
When you put one in the soup.

Marion D. Blyth

*T*he olive has the dubious distinction of being the original—the model—drupe: The Greek word *drýppa* and the Latin, *drūpa*, mean *overripe olive*. Nowadays, *drupe* is the botanical designation for any fruit consisting of an outer skin (epicarp), a fleshy layer (mesocarp), and a hard and woody inner stone (endocarp), which encloses a single seed. The drupe group is a large and important bunch: cherries, prunes, plums, peaches, almonds, and apricots are all drupes. And berries, such as the blackberry and raspberry, are congregations of drupelets. I'll spare my readers the quadrupedantics of speculating on the true meaning of droop-eared (which is usually thus misspelled).

THE BOTANICAL OLIVE

In botanical nomenclature, plants are categorized first by family, then by genus, species, and, finally, by variety. The family to which our edible olive, *Olea europaea*, belongs is Oleaceae, of which the ash, jasmine, lilac, oleander, and privet are also members. Of the genus *Olea*, there are about thirty-five species, some of which are quite spectacular (like the *Olea emarginata*, a large-flowered ornamental that attains heights of sixty feet in India) or exotic (like the *Olea fragrans*, a small, odiferous species from China). None of the other olive species, however, bears an edible fruit. It is only the *Olea europaea*, which has far too many varieties to count, that bears the fruit we press for oil and cure for eating.

The olive is a long-lived and tenacious—one might even say obstinate—plant, which grows in otherwise barren, sometimes arid, places on inhospitable soils of all types. It can live to be a thousand years old. Cut an olive tree down to its roots and from the earth many more will spring. In many localities it need not be irrigated or fertilized, although it is a grateful plant and repays kind treatment with bountiful crops. It does best in climates with long, hot summers, and requires a fair amount of winter chill for fruit set. For that reason it is not grown commercially nearer the equator than 30° north or south latitudes. Contrary to popular belief, it need not be near the ocean to thrive. (Nevertheless, Ernst McCormack swears he can hear the sea in an empty olive jar.)

The olive is an evergreen whose leaves are replaced every two or three years, leaf-fall usually occurring at the same time new growth appears in the spring. The olive's feather-shaped leaves grow opposite one another; their skin is rich in tannin, giving the mature leaf its grey-green appearance.

Olive flowers are small and cream colored, growing on a long stem arising from the axils of the leaves. (The axil is the point where the leaf stem meets the twig.) Because the olive is an evergreen whose leaves hide its flowers, olive-blossom time (usually in late spring or early summer) is not well publicized by the poets. The olive produces two kinds of flowers—a *perfect* flower, containing both male parts (stamens) and female parts (pistils), and a *staminate* flower, with stamens only.

Although most olive varieties are self-pollinating, fruit-set is usually improved by cross-pollination with other varieties. Honey bees are not necessary for cross-pollination if the different varieties are planted fairly closely, for the wind will carry pollen up to a hundred

feet. There are self-incompatible varieties that do not set fruit without other varieties nearby, and there are varieties that are incompatible with certain other varieties. Incompatibility occurs for many reasons—for example, a certain variety of pollen on a certain variety of pistil might be adversely affected by high temperatures.

The olive has won a reputation as an alternate bearer, that is, bearing heavy crops one year, almost no fruit the next. That reputation is ill-deserved and, botanically speaking, untrue. When a very heavy crop is succeeded by a meager one, it is because the tree exhausted its resources so completely one year that it could not produce the next. The problem can invariably be avoided if the tree is pruned carefully every year. Pruning not only regulates production, but it can determine the shape of the tree to accommodate the method of harvesting or the tastes of the gardener. Peasants in Europe used to prune so that a natural ladder was created; multiple trunks can be created by letting either suckers or lower branches grow at the angle at which they are staked. The olive never bears fruit in the same place twice, and usually bears on the previous year's growth. Keeping these factors in mind, farmers can prune for consistent yields year after year.

The dimensions that an olive tree achieves depend on the variety, of course, but to an even greater degree on growing conditions. The largest trees grow in the warmest places, that is, in Tunisia, Morocco, and other North African countries. In California, the Tunisian variety, Barouni, grows no taller than its neighbors, be they Missions or Sevillanos, and comes twenty to thirty feet short of the height of its relatives in Africa, where there are some ninety-foot tall specimens. Climate can so affect the growth habits of a tree that in some cases trees are difficult to identify by variety if they were propagated from cuttings brought from another climate.

It is for that reason the myriad olive varieties cannot be accurately counted; mankind has developed, preserved, and propagated innumerable bearing olive varieties. Some are disease resistant, some are cold resistant, some are drought resistant; some produce small olives with a high oil content, others produce ultra titanics containing little oil. Here are the exotic names of but a few of the major varieties grown in some of the commercial olive-producing countries of the world: Manzanillo, Sevillano, Lechin, Gordal, Picual, Racimal, Argudell, Verdillo, Nevadillo blanco (Spain); Gallego, Verdeal, Cordovil, Carrasquenha, Redondil (Portugal); Frantoio, Moraiolo, Leccino,

Agogia, Rosciolo, Raggia, Nicastrese, Biancolilla (Italy); Caronaiki, Daphnoelia, Mouratolia, Carydolia, Stravolia (Greece); Olivière, Pigalle, Pendoulier, Cailletier, Moiral, Picholine, Lucques (France); Chemlal, Adzeradj, Limli (Algeria); Sam, Girit, Hurma (Turkey); Enduri, R'ghiani, Rasli (Libya); Ouslati, Meski (Tunisia); Ascolano, Sevillano, Mission, Manzanillo, Barouni (United States).

A branch of the olive in flower.

THE HORTICULTURAL OLIVE

Most probably all the varieties of the species *Olea europaea* were developed from the wild olive, or oleaster; consequently, none of the cultivated varieties can be propagated by seed. Seed-propagated trees revert to the original, small-fruited wild variety. Plants grown from seed must be grafted with plant material from the desired variety for the tree to have that variety's characteristics.

Researchers at the turn of the century were quite enthusiastic about grafting fruiting varieties onto rootstock produced from seed, the rootstock of such plants being robust, with a long tap root and symmetrical root system. Such a method of propagation proved to be tedious, however. Even after seed germination was hastened by cracking, or soaking in lye or sulfuric acid, plants grown from seed and grafted after two years seldom bore fruit before they were eight years of age. The method is not used as much today as it was a century ago.

A more commonly practiced method is propagation from cuttings: 12- to 14-inch-long, 1- to 3-inch-wide cuttings from two-year-old wood of a mature tree are planted in a special rooting medium and kept moist. A tree grown from such cuttings need not be grafted to keep the qualities of the cultivated variety, although scientists are finding that combinations of certain varieties can retain the good qualities of both. Furthermore, a tree grown from a cutting bears fruit years before a seed-propagated tree, though not in large quantity.

A third method of propagation is transplanting suckers that grow at the base of mature trees. In most cases, however, those transplants have to be grafted because the suckers grew from the rootstock (and therefore the wild part) of the mature tree.

The variety of an olive tree can be changed by bark-grafting, or top-grafting. Most of the branches of a mature tree are cut off. Quarter- to half-inch-thick grafting wood (or *scions*) from mature trees of the new variety are wedged into the bark of the old tree where the branches were amputated. Usually there are three to four scions per stub. The wounds are covered with grafting wax to prevent infection by disease spores. The tree that has recently undergone this wretched operation is a sad thing to behold, but lo, within three years, the new wood is bearing fruit and all is well for the next few centuries at least.

Experiments with olives are being conducted in all the olive-producing countries—by the United Nations, by national and state governments, and by individual farmers and horticulturists. Carlos Camacho, a Spanish olive grower, is trying a new method at one of his farms: Every few years one trunk of each of his multiple-trunked trees is cut and grafted. Those multi-trunked plants, then, are at once five, ten, and twenty years old. Thus, Carlos's orchards bear consistently year after year, never too young, never peaking, never declining.

The Food and Agriculture Organization of the United Nations, with its network of olive-growing members, is conducting studies with a number of the world's best-producing rootstocks, as well as with wild olive seedlings and seedlings of cultivated varieties. They use a system called "chip budding," where the grafting material is a small bud rather than a branch or twig.

At the Experiment Station in Winters, the University of California is experimenting with grafts of cultivated varieties on rootstocks of plants of different species, but of the same family, Oleaceae. So far the results have been poor; in some cases the grafts have been wholly incompatible and the trees have died.

Fertilizing olive trees with additional supplies of nitrogen has proved quite beneficial. In California, farmers systematically apply fertilizers well ahead of the time flowers develop so the trees can absorb the nitrogen before fruit-set. Some small farmers in the Mediterranean countries are less regular about fertilizing, applying organic fertilizers every couple of years. Other Mediterranean growers (like Maurizio Castelli in Tuscany, who fertilizes twice a year with both organic and commercial fertilizers) use scientifically determined regimes that produce the greatest harvests without damaging the trees or depleting the trees' resources.

In California, irrigation is a necessity because most of the olive orchards are situated where rainfall is unreliable, at best. In many Mediterranean countries, olives are not irrigated, even though harvests sometimes suffer during periods of drought. Because of its small leaves, with their protective cuticle and hairy under-surface that slows transpiration, the olive tree rarely dies during extended dry periods. Nevertheless, with the introduction of new agricultural methods, drip irrigation systems (some of them developed in Israel) are being installed by many Mediterranean olive growers.

Though the olive tree is affected by fewer parasites and diseases than most fruit trees, it does have natural enemies. Around the Mediterranean, the major pests of the olive are what we call the Mediterranean fruit fly, *Ceratitis capitata*, and the olive fruit fly, *Dacus oleae*. A fungal disease called peacock spot is a serious problem wherever olives are grown, affecting leaves and causing defoliation. An even more serious fungal disease for California growers is Verticillium wilt, for which there is no effective treatment save avoiding planting on infested soils and removing damaged trees and branches. A bacterium causes a disease known as olive knot, which is spread by pruning with bacteria-infected tools during rainy months.

Because the olive has fewer pests than other crops, and because the oil in olives retains the odor of chemical treatments, the olive is one of the least sprayed crops of modern agriculture.

GROWING YOUR OWN

I could think of no more beautiful tree to plant in my garden than an olive. So I bought one at my local suburban nursery from a nursery-man who could not let me take the tree home without relating the terrible story of how olive trees had ruined his life.

Many years ago, this lover of plants purchased a little house with an olive-lined drive. Over the next few years he tired of the purple-black stains on his driveway and chopped the trees down. They grew back, only now there were many times more of them. He chopped the trees down and dug out the roots over many months. The trees grew back. He chopped and dug, chopped and dug. And then, as a last resort, he poisoned. The trees grew back. He now has an olive-lined drive.

In this subdued man, however, there is a hint of awe and rever-ence—and even love—for those olive trees, in spite of the damage they did to his self-esteem. The trees ruined his self-image, but they demonstrated a dogged determination to be his companions. His story did not deter me. I have a little grey-green pal up on my clay hillside that I hope will, in ten years' time, be fouling my footpath royally.

Here are some facts. Olives will grow well on almost any well-drained soil up to a pH of 8.5. They will withstand a very hot sum-mer, but temperatures of lower than 15°F will kill the young olive. Olives require full sun. Winter chill is necessary for fruit-set, but lovely ornamentals can be grown where winters are mild.

The varieties commonly available in the United States are Sevil-lano, Ascolano, Barouni, Manzanillo, and Mission (with fruit size descending in that order). If you wish to have no crop, keep in mind that trees designated "nonfruiting" are not always barren; it is safer to prune fruiting wood each season. The Mission is higher in oil content than the other varieties, but home oil production is something to think twice about before trying. The Mission grows tall; the Man-zanillo and Barouni have lower, more spreading growth habits. The

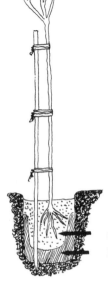

Planting a bare-root olive tree.

native soil mixed with soil conditioner

hole, refilled

Ascolano has a good flesh to pit ratio, and is resistant to olive knot. The Barouni, being a Tunisian variety, withstands extremely high temperatures.

If you plant a young tree, remove a few of the branches at the time of planting. The olive is slow-growing at first, but can be expected eventually to reach thirty feet in height and width in areas with hot summers and fertile soils. For a single trunk, prune suckers and any branches growing below the point where you want branching to begin. For the gnarled effect of several trunks, stake out basal suckers and lower branches at the desired angle.

Some farmers plant trees twenty feet apart, reap harvests until the trees are large, then, when the trees begin to encroach on one another, dig out every other tree, allowing the remaining trees to spread to full size. The trees that are removed do very well when transplanted. If you can find a source for them, you will have twenty years fewer to wait for a large tree. Whether it is young or mature, any newly transplanted tree should be watered frequently; but be careful not to overwater.

Olives can withstand radical pruning. Prune flowering branches in early summer if you wish to prevent fruiting; fruiting can harm a lawn and discolor patios, decks, and walkways. Thin some interior growth to show the olive's branches to best advantage. Cultivate around the base of the tree to prevent competition with weeds.

Do not be impatient with your olive. It's an ill weed that groweth fast—and think of the pleasure it will give your great-grandchildren.

The Nutritive Olive - Fact and Lore

The modern American, with
all his patent contrivances...
will never know... a full
tide of health until he returns
to the proper admixture of
olive oil in his diet.
Until he again recognizes
the value and uses of olive oil,
he will continue to drag
his consumptive-thinned,
liver-shriveled,
mummified-skinned, and
constipated and pessimistic
anatomy about... in a
vain search for lost health.

P. E. Remondino, M.D.*

I would not want to argue against claims that olives are an aphrodisiac—*à chacun son goût*—nevertheless, the cured fruit of the olive tree cannot be construed as health food. The smattering of vitamin E in olives and the iron that some contain** are completely overshadowed by the great quantities of salt added in curing. But the fruit that must be so tortured to be made palatable produces, by the simplest of methods (see chapter entitled "Olive Oil"), the purest and most healthful oil available. Within the pulp of the red-ripe olive are small cellules containing pure, wholesome, quite edible oil which can be extracted simply by pressing. Consequently, the vitamins that other oils lose in chemical extraction and heat (up to 440°F) purification are left unspoiled in cold-pressed, virgin olive oils, which are rich in vitamins E (and alpha-tocopherol), "F," and K, and in betacarotenes.

DIETARY FATS

The subject of fats is a much more complicated one than that of refined versus unrefined foods, however. Whereas it is clear that most unrefined foods are more salubrious than their refined counterparts, there are innumerable theories afoot about the relationship of fats and heart disease, and the matter will not, probably, be resolved for many years. Here is how I, a rabidly subjective arguer for monounsaturates (olive oil being a monounsaturate), have come to view some of the "scientific" evidence. (Note that mine is a layman's view and that new information is continually being published.)

---

*In an essay presented at the Olive Growers' Convention, Sacramento, California 1891.

**Ironically, the ferrous gluconate added to California black olives to fix their color is the same substance used in iron tablets and is quite nutritious.

There are two kinds of dietary fats (fatty acids): saturated and unsaturated. The main sources of saturated fats are animal substances, which also contain cholesterol; the main sources of unsaturated fats are vegetable oils, none of which contains cholesterol in any amount. The unsaturated fats are either monounsaturated (containing oleic acid) or polyunsaturated (containing mostly linoleic acid, with some linolenic and other acids). Olive oil is 73.7 percent monounsaturated fatty acid, 13.5 percent saturated fatty acid, and 8.4 percent polyunsaturated fatty acid (Consumer and Food Economics Institute 1979). Because it is a vegetable product, olive oil contains no cholesterol.

The most popularly accepted of the theories about the dietary causes of atherosclerosis (the buildup of placque in arterial walls) is that high serum (blood) cholesterol levels contribute to the buildup, that saturated fats in the diet raise serum cholesterol levels, and that polyunsaturates, among other things, deplete serum cholesterol levels. It is because of this theory that many people restrict their consumption of eggs and meats, and cook with polyunsaturated fats such as safflower, sunflower, and corn oils.

Most researchers do not question the link between serum cholesterol and coronary disease; there is ample evidence to show a direct relationship (although new studies are raising questions about other substances, such as hormones). But the reasons certain persons develop high serum cholesterol levels are not fully understood, and that is where much controversy has arisen.

Some statistics seem to contradict the theory that there is a direct link between saturated fat intake and serum cholesterol levels. While the consumption of saturated fats in America over the past seventy years has remained constant, the consumption of what has been assumed to be healthful—polyunsaturated fats—has doubled. And the incidence of coronary disease has gone up enormously in the past century, although quite recently there has been some evidence to indicate that it may have begun to decline. Some scientists maintain that the cause is not, therefore, saturated fats in the diet. Others go even farther and state that refined carbohydrates such as white flour and sugar are to blame, for the consumption of refined carbohydrates increased at about the same time as that of the polyunsaturates.

A survey by Dr. L. Michaels supports the theory that sugar contributes to heart disease. During the past two centuries in England, the consumption of saturated fats remained the same, the consumption of sugar rose over twenty-five-fold, and the incidence of angina pectoris (one symptom of heart disease) rose astronomically (Michaels 1966).

To further complicate the picture, it appears that, although polyunsaturates may reduce serum cholesterol, most refined polyunsaturated oils are low in an antioxidizing constituent of vitamin E—alpha-tocopherol—because it is destroyed by processing. The absence of the constituent is conjectured to permit the process of lipid peroxidase in the body, which may produce toxins that contribute to various diseases, including cancer. Polyunsaturates may be reducing heart disease by lowering serum cholesterol levels, but at the same time they may be contributing to other, equally life-threatening diseases. Because cold-pressed olive oil is produced without heat, none of its alpha-tocopherol is lost (Dinaburg and Akel 1976).

Although the questions regarding saturated fats, polyunsaturated fats, and serum cholesterol have yet to be answered to the satisfaction of all, much has been written and many theories have been forwarded. Monounsaturates, on the other hand, are rarely seen by the American public as being either the culprit that saturated fats have become or as the health-giving agent that polyunsaturates are commonly considered to be. American scientists have paid monounsaturates little attention, perhaps because most Americans consume so little olive oil, the world's chief source of monounsaturated oil.

A new study being conducted with Greek, Greek-American, and American boys of non-Greek descent is breaking new ground and producing some interesting results. The Greek-American boys living in the United States and eating American diets rich in polyunsaturated and animal fats develop nearly the same high serum cholesterol levels, at about the same age, that non-Greek American boys do eating the same diets. The serum cholesterol levels of Greek boys in Greece, however, who consume 15 to 30 percent of their calories in olive oil, remain much lower.

Even more interesting is the effect of diet on the fatty tissues of the subjects. The boys in Greece have high levels of monounsaturated acids in their adipose tissues, whereas the boys in the United States have high levels of polyunsaturated and saturated acids in the same tissues. The doctors conducting the tests have conjectured that the Greek subjects' diets (which are rich in olive oil) may help create a "biochemical milieu" that leads to lower serum cholesterol levels and to lower coronary disease rates in Greece. In the United States, there is ten times the incidence of coronary disease among men between the ages of 55 and 65 than in Greece (*Medical World News* 1982).

On the other hand, Jason Peacock's old saying, "You got a theory? I'll find you a study," is good to bear in mind. We may have to accept the fact that we moderns have not yet succeeded in establishing to the satisfaction of Science those substances that constitute a healthy diet—and an effective arsenal of medicines.

THE CURATIVE OLIVE

Other generations and civilizations thought they knew instinctively what was good for them. They followed on faith (and superstition) the recipes for medicaments that were handed down to them. Olive oil, and other products and parts of the olive tree such as leaves and branches and gum from the trunk, are among those natural ingredients for remedies that have been around for hundreds or thousands of years.

The first use for olive oil was as an anointment, it is generally believed, not as food or fuel: It was a soothing lubricant and emollient, and its application was so important it became a religious rite. In the Egyptian Ebers Papyrus (a medical text dating from 1550 B.C.), olive oil is listed among 700 therapeutics, and was taken orally as well as being rubbed into the skin.

As long as olive oil has been used as a medicine in and of itself, it has also been used to convey other curatives. And its use is as current as this year's *U.S. Pharmocopeia*. Olive oil is useful for its ability to penetrate into the follicles, to protect against infection, and to protect wounded or irritated internal or external tissues. It is even injected intramuscularly, to convey medicines, because it is so easily assimilated.

I have heard well-educated persons in modern, enlightened Italy proclaim olive oil a preventive for cancer, arteriosclerosis, embolisms, and thrombosis. Those beliefs may well be based on medical practice from a previous century, on personal experience, or on the long-abiding grasp of superstition and folklore. Here is a list of ailments for which olive oil is an effective treatment, compiled from Mediterranean folklore and from nineteenth-century American medical practice:

— Intestinal parasites: "In cases of tapeworm," said Dr. Remondino a hundred years ago, "olive oil often carried the gentleman, head, tail, and body, simply by its weight and volume."

— Bladder and kidney afflictions

— Inflammations of mucous membranes

— Simple diarrhea, dysentery, colic, flatulence, and constipation

— The discomfort of teething

— Rheumatic diseases

— Yellow fever

— Tumors and diseases of the throat glands

— Any disease resulting from mal-assimilation of nutrients

Olive oil, and all parts of the olive tree, figure importantly in potions for all sorts of ills. Here are some useful recipes:

— Olive oil liberally mixed with powdered charcoal for mushroom poisoning. Or mixed with crushed seeds of mimosa against any poison.

— Olive oil mixed with an equal part limewater for burns.

— Red clay mixed with half olive oil, half water to make a paste for suppurating pimples.

— To every quart of olive oil add two hundred legs of centipedes, a piece of snake's skin, and the sprouts of flowered ruda (a Spanish herb). Boil it until it is reduced a third. Good against paralysis.

— Leave water that has been used to marinate olives seven times (once each year for seven years), and to which salt and fennel have been added, in the open sun and rain for forty days. Cures venereal disease.

— In a small bottle mix olive oil, camomile flowers, anise, and a few cloves of garlic, finely chopped. Let it sit nine days. Shake before rubbing on chest, back, or the soles of feet for coughs, aches, and soreness, respectively.

— There is a plant in Spain, the *calabruixa*, which witches like to keep for themselves. But if a friend or relative is under a witch's spell, find a *calabruixa*, mash it in olive oil, and leave it uncovered in the open air in a place hidden from witches. The potion will break the spell.

— Olive oil mixed with betony (a kind of mint) water prevents drunkenness.

— In the Levant, olive oil is rubbed on the bodies of babies and the sick to relax them.

— A drop of warm oil relieves earaches.

— Olive oil, rubbed into the hair and scalp, encourages luxuriant growth and makes hair shiny.

— A mixture of burnt garlic and olive oil cures women's illnesses and ulcerations of the head.

— A soap made with olive oil and burned rose laurel (or sometimes sulfur) cures skin diseases.

— In Iran, the wild olive (prepared I know not how) is a cure for hemorrhoids.

OLIVE LORE

Let us go beyond the medicinal now and say that olive oil, olives, olive leaves, and olive branches appear in folklore throughout the Mediterranean to ward off all manner of evils from witches to overactive libidos.

— Venetians say an olive branch on the chimney piece wards off lightning.

— Throughout Italy, a branch over the door keeps out witches and wizards.

— In Spain, an olive branch makes the husband faithful and the wife master of the house.

— In Bilda, Algeria, there was an ancient olive tree into which the infirm drove nails to cure their ailments.

— In Lebanon, during times of famine when only bread and oil were available, parents consoled their children by saying, "Eat bread with oil and hit your head on the wall." That means that such a diet will make you so strong you can take any kind of abuse.

— In Lebanon "He seasoned his greens with his own oil" is a byword for self-sufficiency. There, "wheat and oil are the pillars of the house," and olives are "the sultan of the table."

— In Iran, talismans are washed in olive oil. Sometimes oil is solemnly burned, and frequently the formulae that appear on the talismans are written with olive oil.

— In Morocco, to enter someone's house carrying olive oil and to leave without giving some away brings a curse.

Double, double oil and trouble ...

— In Jordan, the morning after a funeral and for the next three Thursdays (or Wednesdays, if the deceased was murdered) meals are served consisting mainly of various kinds of pancakes fried in olive oil.

Folklore prescribes rules pertaining to the olive tree and its precious harvest. There is the Spanish belief that an olive tree touched by a prostitute will be unfruitful; likewise the olive trees of an unfaithful husband. Some believe the fruitfulness of the olive tree will increase if it is tended by young and innocent children. It is said that the quality of the olive crop is predicted by the colors in the rainbow.

There is a Moroccan legend about the pious olive tree in the Garden of Eden on each of whose leaves the name of God is invisibly written. The olive tree advised Eve in the garden not to listen to the serpent, but she ignored the advice. When Adam and Eve ate of the forbidden fruit and discovered their nakedness, the olive tree again advised them, "Repent. God will forgive you." But Adam replied that he was afraid and wanted to hide. The olive said, "Go, then, for no one can stop a man when he walks to his doom." To reward the olive tree for its love of mankind, God said, "Your leaves will be the color of silver and emerald. Your fruit will be worth its weight in gold. Flocks will find shelter under your leaves, and saints will seek your shade and bless you."

There is another legend, about the Church of Nossa Senhora da Oliveira (Our Lady of the Olive Tree) at Guimaraens in Portugal. During the Gothic occupation, the future king Wamba was herding cattle at that spot when he was told that he had been chosen king of the Goths. He thrust his staff into the ground and swore that he would not be king until that stick of olive wood grew leaves. God demonstrated his will by causing the stick to send forth not only leaves but branches and fruit, and Wamba agreed to be king. (A similar story of an olive stick that sends out leaves appears in the Greek legend of Jason and Medea—testimony to the faith people have, perhaps, in that tree's ability to survive death.)

At the very least, the numerous legends, superstitions, and potions based on olives or olive oil attest to the olive's importance through the ages. Let us end this chapter with the conservative (if boring) description of olive oil from *The Dispensatory of the U.S.A.* If olive oil is nothing else—putting what tradition, doctors, and health enthusiasts advise aside—it is ". . . nutritious and mildly laxative. Usual dose as laxative or cholagogue is 15 to 60 ml (½ to 2 fluid ounces). Like other fats it delays gastric emptying. Causes contraction of the gall bladder and is used for diagnostic purposes. In atonic conditions of the gall bladder it is used therapeutically, although, at first, it may cause discomfort. In the form of enema (150 to 500 ml warmed to body temperature) it is a useful remedy in fecal impaction. As an emollient it is applied topically to the skin or mucous membranes."

The Olive
in Art
and Literature

If I could paint and had
the necessary time, I should
devote myself for a few years
to making pictures only
of olive trees.

Aldous Huxley
The Olive Tree

$\mathcal{G}$rowing on the fringes of the desertlike *garrigue* of Languedoc, up the steep cliffs of the Grecian islands, in fruitful masses along the plains outside Seville, and as silent fortresses over the oases in the North African deserts, the olive, with its grey, venerable presence, is more symbolic of the Mediterranean than any other tree—even the cypress, even the poplar. It is the tree of western art and literature, for it is at once the universal symbol of peace, the most important and useful tree to the Mediterranean economy (for all ages, including the present), and of inestimable beauty. Poets have personified it much as they have the ancient oak, for it has weathered the adversities of centuries and stands able, it would seem, to outlast the vacuity of modern existence. It is a nurturing tree, with its yield able to satisfy all manner of needs—the needs for nutrient, curatives, solace, and light. As nurturing as the olive is, however, it is not a feminine or maternal tree. Its thick, gouged trunk makes it clumsy beside the slim, graceful cypress; its grey, rather sparse foliage lacks the voluptuousness of the bright green elm. The genderless olive is quite simply human—taking on a tortured, hoary aspect against a grey-black sky, or conveying a calm, earthbound happiness beneath a blue one.

IN LITERATURE

The olive was tamed from its wild, barren state before western man discovered his God, and before his three great modern religions took form. Humanism and the olive came into being concurrently. This lugubrious tree produced for the Hebrews an oil that was a symbol of joy and plenty; the Old Testament makes frequent mention of it. Although for the Greeks the olive meant victory (their wild olive wreaths are described by Ruskin as "cool to the tired brow... type of grey honour") and was a symbol of supplication (as described in the opening of *Oedipus Rex*), it is as a symbol of peace that it has survived thousands of years into modern times. Men who have never seen an olive leaf recognize it immediately in cartoons, on the flag of the United Nations, in the beak of Picasso's dove.

The olive first appears in the literature of the English language as a symbol of peace. Chaucer made a list of trees for his fourteenth-century readers, and among them, of course, was "The olyve of pees." Shakespeare, too, knew it as an emblem of peace, and used it freely in his plays. It appears in *Henry VI Part III*: "To whom the heavens in thy nativity/Adjudged an olive branch and laurel crown/ As likely to be blest in peace and war"; and in *Antony and Cleopatra*: "The three nook'd world shall bear the olive freely." Two lovely lines from Sonnet 107 read: "Incertainties now crown themselves assured, And peace proclaims olives of endless age." Milton's *Paradise Lost* makes reference to the olive, of course, when the dove (from Genesis) brings Noah the good news: "An Olive leaf he brings, pacific signe." Alexander Pope uses it allusively as well: "Where Peace descending bids her olives spring."

Milton speaks of the olive not only as a symbol of peace but as an important tree that flourished at the height of ancient Greece. He writes of "The olive groves of Academe/Plato's retirement, where the Attic bird/Trills her thick-warbled notes the summer long." But it was the Romantic poets who appreciated the olive for its beauty. In 1791, Cowper referred to "the luxuriant olive." Byron, well traveled on the shores of the Mediterranean, refers to the land "Where the citron and olive are fairest of fruit." Others ascribed human traits to it by using pathetic fallacy. Lamartine, appreciative in part, perhaps, because the olive grew on his native French soil, wrote in 1839, "It was those very olives themselves, the venerable witnesses of so many days, written on earth and in heaven."

However, no one loved the Mediterranean landscape more than the expatriate writers—those who had left their countries and lived for decades in their "new" lands. Twentieth-century writers from England and America flocked to the Mediterranean in the '20s and '30s, for numerous and complex reasons. They were attracted not only by the social, political, and intellectual climate of the region, but by the Mediterranean countryside and its singular aspect as well. And what would that countryside be without the olive groves?

Aldous Huxley was so taken with the olive and its airy clarity that he wrote an essay about it entitled "The Olive Tree" and called his collection of essays written in the '20s ands '30s by the same name.

Ralph Bates, an English expatriate who arrived in Spain in 1923 and became a political activist in the '30s, wrote *The Olive Field*—a book that takes place in an Andalusian village (Los Olivares) named for the olive groves that surrounded it at the time of the Spanish revolution. What better backdrop for a story of turmoil, nobility of spirit, and permanence and transitoriness than the terraces of olive trees in a parched Spain, where Bates's characters harvest and tend their landlord's olives. The descriptions of olive trees and groves are lovingly wrought, and the olives become a perfect metaphorical device.

— Mon vieux, ta peinture manque de chaleur.
— Et la peintre donc!
(After a lithograph by Honoré Daumier.)

IN ART

The painters for whom the Mediterranean landscape seems to have been created were the Impressionists, who best captured the spirit of southern France and its tree, the olive. And how differently they expressed it. Renoir could not bear to paint a grey tree. He painted gold and, more exuberantly, pink, olive trees. Matisse, though a more abstract painter perhaps, preferred the olive's natural hue and painted grey olives at Collioure and Corsica. Cezanne, while loving his black cypresses and chestnuts as well, found a place for the grey-green olive in his rich landscapes. Bonnard painted the olive grey, but the grey did nothing to subdue the cheerful aspect of his St. Tropez. And the same could be said of Derain, a little later, who was able to capture the landscape of the Mediterranean with just a few, almost primary, colors.

It was van Gogh, however, who was most fascinated by the olive. And who could capture it better? He wrote from Provence: "I struggle to apprehend this. It is silver, perhaps a little blue, or perhaps somehow green—whitish bronze over reddish ochre earth. It is very difficult, very difficult. Even so, it attracts me. I like to contrast these colors with silver and gold. Someday I will obtain a personal impression as I did with sunflowers." He painted nineteen pictures of olive trees. "Olive Trees—Yellow Sky with Sun"; "Olive Trees—Bright Blue Sky"; "Olive Trees—Orange Sky"; "Olive Trees—Pink Sky." How full of movement his silver-green trees are. Their elbows are gnarled, their roots are embedded in an undulating yellow earth, their leaves explode into the sky. The olive trees van Gogh painted when he was at the sanitarium in St. Remy are sad, seething monuments to his disturbed life. Those turbulent paintings notwithstanding, van Gogh did paint a happier scene with olives: a couple plucking olives from young trees in an "Olive Yard."

The olive tree has never had a visual presence in American art. Americans do not know or appreciate the tree, even in California. But landscape architects are beginning to incorporate it into their designs, and some day, perhaps, it will become commonplace here.

BIBLIOGRAPHY

Adams, Catherine F. 1975. *Nutritive Value of American Foods*. Agricultural Handbook No. 456. Washington, D.C.: USDA.

Aguilar, Jeannette. 1966. *The Classic Cooking of Spain*. New York: Holt, Rinehart and Winston.

Antas, Mohamed A., Margaret Olson, and Robert E. Hodges. 1964. "Changes in Retail Market Food Supplies in the U.S. in the Last 70 Years in Relation to the Incidence of Coronary Heart Disease with Special Reference to the Dietary Carbohydrate and Essential Fatty Acids," *American Journal of Clinical Nutrition* 14 (March).

Barich, Bill. 1983. "Tuscan Spring." *The New Yorker* (May 30).

Bates, Ralph. [1936] 1966. *The Olive Field*. New York: Washington Square Press.

Bensoussan, Meir, and Gabriel Grabi. 1960. *A Survey on Methods of Picking Olives for Pickling and for Oil*. Tel-Aviv: Institute of Productivity.

Bugialli, Guiliano. 1982. *Classic Techniques of Italian Cooking*. New York: Simon and Schuster.

————— *The Fine Art of Italian Cooking*. 1977. New York: The New York Times Book Company, Inc.

California State Board of Horticulture. 1900. *Investigation Made by the State Board of Horticulture of the California Olive Industry*. Sacramento: State Printing Department.

Center for Science and Public Interest. 1982. "Chemical Cuisine Chart." Washington, D.C.

Chimenti, Elisa. 1965. *Tales and Legends of Morocco*. Translated by Arnon Benamy. New York: Ivan Obolensky.

Ciurana, Jaume, and Llorenc Torrado. 1981. *Els Olis de Catalunya i la Seva Cuina*. Barcelona: Department d'Agricultura.

Consumer and Food Economics Institute. 1979. *Composition of Foods: Fats and Oils*. Agricultural Handbook No. 8–4. Washington, D.C.: U.S. Government Printing Office.

Coutance, A. 1877. *L'Olivier*. J. Rothschild, ed. Paris.

Cronin, Isaac, Jay Harlow, and Paul Johnson. 1983. *The California Seafood Cookbook*. Berkeley: Aris Books.

David, Elizabeth. 1950. *A Book of Mediterranean Food*. Harmondsworth: Penguin Books, Inc.

————— 1960. *French Provincial Cooking*. New York: Penguin Books.

Dean and DeLuca Imports Incoporated. 1982. *March 1982 Newsletter*. New York: Dean and DeLuca Imports Incorporated.

Dinaburg, Kathy, and D'Ann Ausherman Akel. 1976. *Nutrition Survival Kit*. San Francisco: Panjandrum Press/MidPress Productions.

The Dispensatory of the U.S.A. 25th ed. 1955. Philadelphia: J. B. Lippincott Company.

Donaldson, Dean, William A. Dost, and Richard Standiford. 1983. *Heating Your Home with Wood*. Agricultural Sciences Publications Leaflet 21336. Berkeley: University of California.

Edwards, I. E. S., C. J. Gadd, N. G. L. Hammond, and E. Sollberger. 1975. *The Cambridge Ancient History* 3rd ed. Cambridge: Cambridge University Press.

Fitch, Cleo Rickman. 1982. "The Lamps of Cosa," *Scientific American*, vol. 247, no. 6 (December).

Flamant, Adolphe. 1887. *A Practical Treatise on Olive Culture*. San Francisco: Louis Grégoire and Company.

Food and Agriculture Organization of the United Nations. 1976. *Report of the Third Session of the FAO Olive Production Committee, Khania, Greece*. Rome: F.A.O.

Friedman, Nancy. 1982. "A Short History of the California Olive," *San Francisco Magazine* (December).

Anonymous. 1983. "Oils and Vinegar," *The Gourmet Retailer* vol. 4, no. 2 (February). No. Miami: Edward Loeb.

Harris, Lloyd J. [1974] 1979. *The Book of Garlic*. Berkeley: Aris Books.

Hartmann, H. T., K. W. Opitz, and J. A. Beutel. 1980. *Olive Production in California*. Agricultural Sciences Publications Leaflet 2474. Berkeley: University of California.

Hazan, Marcella. 1982. *The Classic Italian Cook Book*. New York: Alfred A. Knopf.
———— 1978. *More Classic Italian Cooking*. New York: Alfred A. Knopf.

Hilgard, E. W. and P. C. Remondino, eds. 1891. *Proceedings of the Olive Growers Convention, July 8, 1891*. Sacramento: California State Printing Office.

Hodgson, Moira. 1981. "A Sampling of the World's Olive Oil," *The New York Times* (September 2).

Huxley, Aldous. 1937. *The Olive Tree*. New York: Harper and Brothers Publishers.

Jacobson, Michael F. 1975. *Nutrition Scoreboard—Your Guide to Better Eating*. New York: Avon Books.

Johnstone, Mireille. 1976. *The Cuisine of the Sun*. New York: Random House.

Lelong, B. M. 1889. *The Olive in California*. Sacramento: State Board of Horticulture.

Majnoni, Francesco. 1981. *La Badia a Coltibuono—Storia di una proprietà*. Firenze: Francesco Papafava.

Marvin, Arthur Tappan. *The Olive—Its Culture in Theory and Practice*. 1889. San Francisco: Payot, Upham and Company.

Medical World News. 1982. "Why Greece's coronary rate is low: lots of olive oil?" (March 15).

Michaels, L. 1966. "Aetiology of Coronary Heart Disease: An Historical Approach," *British Heart Journal* 28.

Montagné, Prosper. 1961. *Larousse Gastronomique*. New York: Crown Publishers, Inc.

Morettini, Alessandro. 1950. *Olivicoltura*. Rome: Ramo Editoriale degli Agricoltori.

The New Encyclopedia Britannica, Macropaedia, 15th ed. 1973–1974. "Oils." Chicago: Helen Hemingway Benton.

Nieto, J. Mìguel Ortega. 1955. *Las Variedades de Olivo Cultivadas en España*. Madrid: Estacion de Olivicultura de Jaen.

Olney, Richard. 1975. *Simple French Food*. New York: Atheneum.

Pect, Harry Thurston, ed. 1965. *Harper's Dictionary of Classical Literature and Antiquities*. New York: Cooper Square Publishers, Inc.

Pohndorff, F. 1884. *A Memoir on Olive Growing*. San Francisco: California State Horticultural Society.

Roberts, J. M. 1976. *History of the World*. New York: Alfred A. Knopf.

Root, Waverly. 1968. *The Cooking of Italy.* New York: Time-Life Books, Inc.

——— 1971. *The Food of Italy.* New York: Random House, Inc.

Shakespeare, William. 1948. *The Complete Works.* G. B. Harrison, ed. New York: Harcourt, Brace and Company.

Smith, William, William Wayte, and G. E. Marindin, eds. 1901. *A Dictionary of Greek and Roman Antiquities.* London: John Murray.

Standish, Robert. 1961. *The First of Trees—The Story of the Olive.* London: Phoenix House.

Sunset Magazine and Sunset Books, eds. 1967. *Sunset Western Garden Book.* Menlo Park: Lane Publishing Co.

Taylor, Frank. 1954. "California's Strangest Crop," *Saturday Evening Post* (October 2).

Waters, Alice. 1982. *The Chez Panisse Menu Cookbook.* New York: Random House.

Anonymous. 1982. "Why Greece's coronary rate is low: lots of olive oil?" *Medical World News* (March 15).

Wolfert, Paula. 1973. *Couscous and Other Good Food from Morocco.* New York: Harper and Row, Publishers.

York, George, 1979. *ABC's of Home-Cured, Green-Ripe Olives.* Agricultural Sciences Publications Leaflet 21131. Berkeley: University of California.

Zyw, Leslie. 1981. "Tuscan Cold Pressed Extra Virgin Olive Oil," *Petits Propos Culinaires 7.* London: Prospect Books (March).

SELECTED OLIVE OIL RESOURCE LIST

This is a regional guide to selected shops that stock quality olive oils. If you cannot find quality oils in your area, write Aris Books, 1635 Channing Way, Berkeley, California 94703 and you will receive a mail order catalog.

California, Northern
Oakville Grocery, 1555 Pacific Ave., San Francisco 94109
Say Cheese, 856 Cole St., San Francisco 94117
Williams-Sonoma, 576 Sutter St., San Francisco 94102
The Kitchen, 2213 Shattuck Ave., Berkeley 94704
Made to Order, 1576 Hopkins St., Berkeley 94707
Narsai's Market, 385 Colusa Ave., Kensington 94707
Ratto's, 821 Washington, Oakland 94607
The Pasta Shop, 5940 College Ave., Oakland 94618
Bon Appetit, 1599 Tiburon Blvd., Tiburon 94920
Rossetti's, 520 San Anselmo Ave., San Anselmo 94960
Corti Brothers, 5770 Freeport Blvd., Sacramento 95822
William Glen, 2651 El Paseo Lane, Sacramento 95812
Traverso Gourmet Food, 3rd Ave. and B St., Santa Rosa 95401
Williams-Sonoma, Vallco Fashion Park, 10123 N. Wolfe Rd., Cupertino 95014;
    36 Town & Country Village, Palo Alto 94301
Cosentino's Market, 2666 S. Bascom Ave., San Jose 95124

California, Southern
Durham's 243 S. Muirfield Rd., Los Angeles 90004
Faire La Cuisine, 8112 Melrose Ave., Los Angeles 90046
Hugo's Fine Meats, 8401 Santa Monica Blvd., Los Angeles 90069
Irvine Ranch Farmer's Market, 142 San Vicente, Los Angeles 90048
Pasta, Pasta, Pasta, 8160 W. 3rd St., Los Angeles 90064
The Wine Shop, 223 N. Larchmont Blvd., Los Angeles 90004
Two's Company, 6009 Waring Ave., Los Angeles 90038
Williams-Sonoma, Beverly Center, 131 N. La Cienega Blvd., Los Angeles 90048;
    South Coast Plaza, 3333 Bristol St., Costa Mesa 92626; 438 N. Rodeo Dr.,
    Beverly Hills 90210; The Commons, 146 S. Lake Ave., Pasadena 91101
Montana Mercantile, 1324 Montana Ave., Santa Monica 90403
Tutto Italia, 8657 Sunset Blvd., West Hollywood 90210
Pierre Lafond and Co., 516 San Ysidro Rd., Santa Barbara 93108
D. Crosby Ross, 1117 State St., Santa Barbara, 93101
Perfect Pan, 4044 Gold Finch, San Diego 92103
Vittorio's, 1302 Camino Del Mar, Del Mar 92014

Colorado
Williams-Sonoma, 1460 Larimer St., Larimer Sq., Denver 80202

Connecticut
Hayday Market, 1026 Post Road East, Westport 06880
The Good Food Store, 856 Post Road, Darien 06820

Florida
Brinkley's Market Place, 354 S. County Rd., Palm Beach 33480

Georgia
Proof of the Pudding, 980 Piedmont, Atlanta 30309

Illinois
Mitchell Cobey Cuisine, 100 E. Walton, Chicago 60611
Treasure Island Foods, 3460 N. Broadway, Chicago 60657

Kentucky
Lotsa Pasta, 2424 Bardstown Rd., Louisville 40205

Maine
Winemporium, Highland Mill Mall, Camden 04843

Maryland
Gourmet, 3713 Old Court Rd., Baltimore 21208

Massachusetts
Cardullo's, 6 Brattle St., Cambridge 02138
Williams-Sonoma, Copley Place, 100 Huntington Ave., Boston 02116

Michigan
City Delights, 516 Brush St., Detroit 48226
Zingerman's Deli, 422 Detroit St., Ann Arbor 48104

Minnesota
Crocus Hill Grocery, 674 Grand Ave., St. Paul 55105
Epicure's Market in the Galleria, 3600 W. 70th St., Edina 55435
Williams-Sonoma at Harold, 818 Nicollet Mall, Minneapolis 55402

Missouri
Pasta Presto, 4105 Pennsylvania, Kansas City 64111
The Pampered Pantry, 8139 Maryland Ave., St. Louis 63105

New Jersey
Williams-Sonoma, The Mall at Short Hills, Short Hills 07078

New York

Balducci's, 424 6th Ave., New York City 10011
Bloomingdales, 100 3rd Ave., New York City 10002
Dean and DeLuca, 121 Prince St., New York City 10012
Incredible Edibles, 857 2nd Ave., New York City 10017
Zabars, 2245 Broadway, New York City 10024

Ohio

Pastiche, 8699 Chagrin Blvd., Cleveland 44122

Oklahoma

Creative Cookery, 6509 North May, Oklahoma City 73116

Oregon

Panache, 5331 SW Macadam, Space 150, Portland 97201

Pennsylvania

Chestnut Hill Cheese Shop, 8509 Germantown Ave., Philadelphia 19118
Shady Side Market, 5414 Walnut St., Pittsburgh 15232

Texas

Di Palma's, 1520 Lower Greenville Ave., Dallas 75206
Sakowitz, Pacemaker Shop, P.O. Box 1387, Houston 77001
Williams-Sonoma, Highland Park, 51 Highland Park Bldg., Dallas 75205

Vermont

La Bottega, 134 Church Street, Burlington 05401

Washington

Frederick & Nelson, 5th & Pine, Seattle 98111

Washington, D.C.

AA Liquors, 1909 Pennsylvania Ave. NW 20006
Food & Company, 1200 New Hampshire Ave. NW 20036

Wisconsin

Sendik's, 4027 Oakland, Milwaukee 53211
Stock Pot, 6119 Odana Rd., Madison 53719

Index

COOKBOOKS FROM ARIS BOOKS

The Book of Garlic by Lloyd J. Harris. The book that started America's love affair with garlic. It consolidates recipes, lore, history, medicinal concoctions and much more. "Admirably researched and well written."—Craig Claiborne in *The New York Times*. Third, revised edition: 286 pages, paper $9.95

The International Squid Cookbook by Isaac Cronin. A charming collection of recipes, curiosities and culinary information. "A culinary myopia for squid lovers."—*New York Magazine*. 96 pages, paper $6.95

Mythology and Meatballs: A Greek Island Diary/Cookbook by Daniel Spoerri. A marvelous, magical travel/gastronomic diary with fascinating recipes, anecdotes, mythologies, and much more. ". . . a work to be savored in the reading. . ."—*Newsweek*. 238 pages, cloth $14.95

The California Seafood Cookbook by Isaac Cronin, Jay Harlow and Paul Johnson. The definitive recipe and reference guide to fish and shellfish of the Pacific. It includes 150 recipes, magnificent fish illustrations, important information and more. ". . . one of the best manuals I have ever read. . ."—M. F. K. Fisher. 288 pages, cloth $18.95, paper $10.95

The Art of Filo Cookbook by Marti Sousanis. International entrées, appetizers and desserts wrapped in flaky pastry. 144 pages, paper $8.95

To receive the above titles, send a check or money order made out to Aris Books for the amount of the book plus $1.25 postage and handling for the first title, and 75¢ for each additional title. To receive our current catalogue of new titles, send your name and address plus 50¢ for postage and handling.

Aris Books, 1635 Channing Way, Berkeley, CA 94703 (415) 843-0330